Certifiable Optician

THE BIG BOOK

Visualize and Understand

Need-to-Know Concepts

ABO EXAM 2026 - 2029

by D. R. Armer, LDO, MBA

PASS THE ABO THE FIRST TIME!

Over 200 Illustrations to Make Learning Easy!

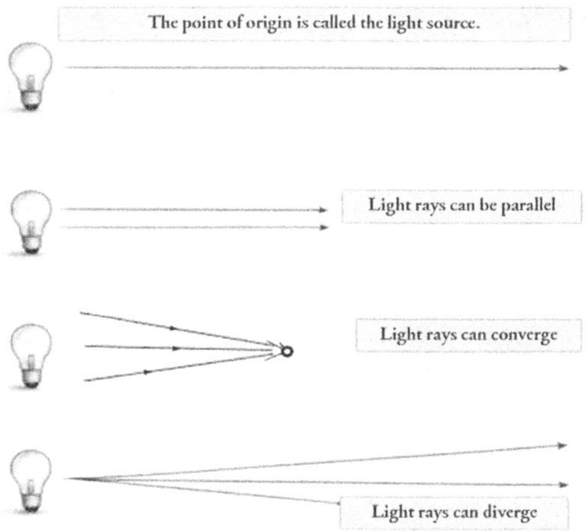

Color this image.

COLORING PAGES &

PLAIN LANGUAGE

Prism Direction Rules
Splitting

Prism may typically be split for cosmetic reasons. Placing all the prism in one lens may make a very thick heavy lens on one side. Split the prism according to the direction rules to balance the prism.

Turn this:

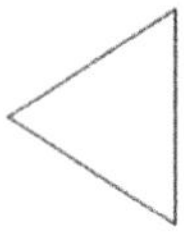

Into this:

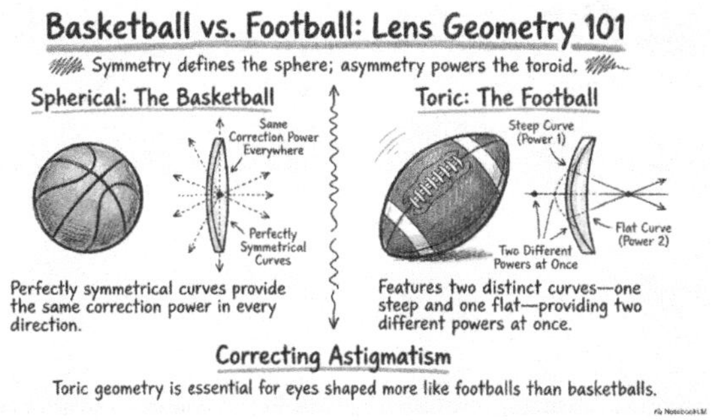

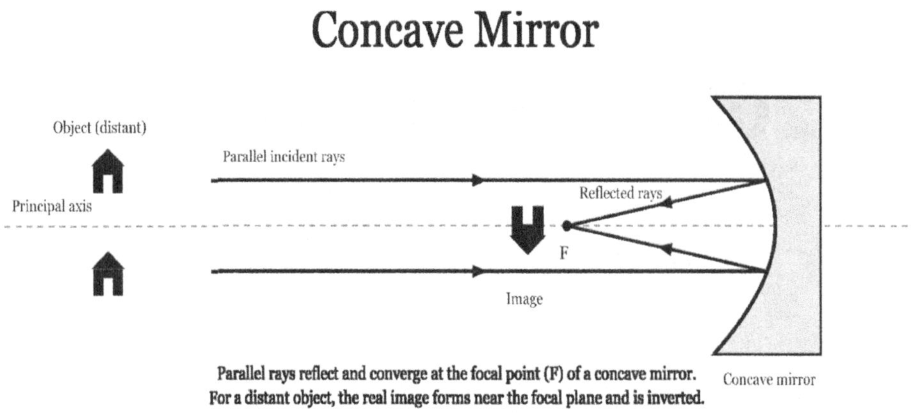

Certifiable Optician, THE BIG BOOK, covers topics in a way that helps you visualize key topics

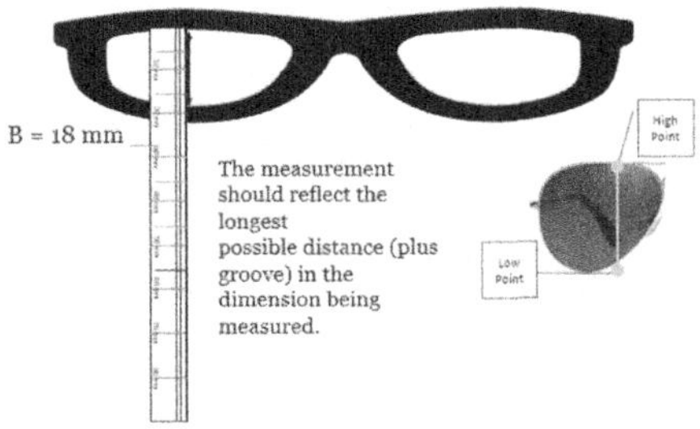

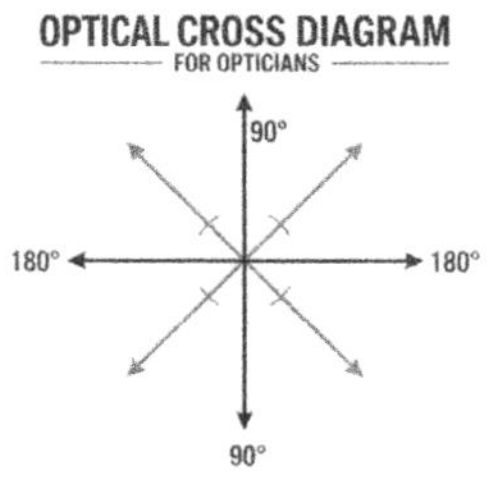

This is the most efficient method that *places great value on your time.* No fluff, just facts.

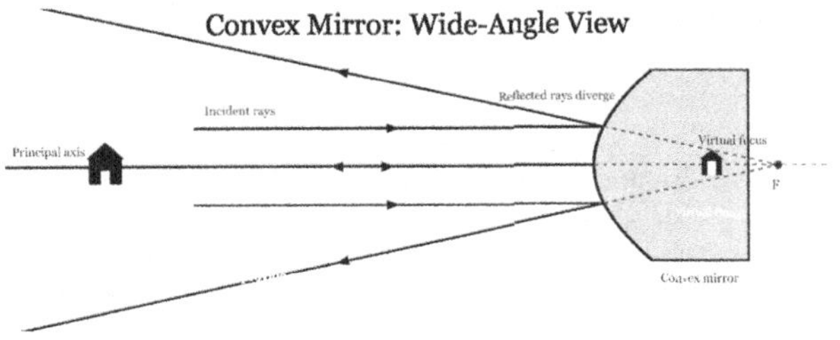

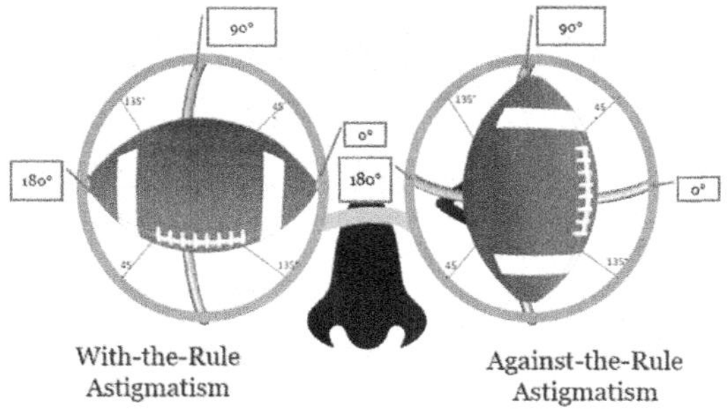

D. R. Armer

armerclad1066@gmail.com

Quantity sales. Special discounts are available on quantity purchases by corporations, associations, and others. For details, contact the publisher at the address above.

Printed in the United States of America

First Edition: April 2026

ISBN: 978-1-7350164-3-6

The information provided in this book is for educational and informational purposes only and is not intended as medical advice, diagnosis, or treatment. The author is not a licensed physician or healthcare professional. Always seek the advice of your physician or other qualified health provider with any questions you may have regarding a medical condition.

Never disregard professional medical advice or delay in seeking it because of something you have read in this book. If you think you may have a medical emergency, call your doctor or 911 immediately. Reliance on any information provided in this book is solely at your own risk. The author and publisher shall have neither liability nor responsibility to any person or entity with respect to any loss or damage caused or alleged to be caused directly or indirectly by the information contained in this work.

Most graphics were created line by line in MS PowerPoint. Learning was enhanced in some cases by creating additional graphics via iterative, sequential prompting into AI tools like Gemini, NotebookLM and ChatGPT. These tools were used primarily for brainstorming and reference sourcing. The prompting that engineered any AI output was original work by the author.

Table of Contents

Testing, Testing

Blueprint for Efficient Learning

"Our main point was to emphasize that practice recalling knowledge and skills is at least as important as how students are initially exposed to information." (Yang et al., 2019, p. 1)

THE study referenced above, along with literally dozens of other studies, have confirmed that the process of practice testing is the most effective learning tool. Hands down, it is better than re-listening to lectures, watching videos, or highlighting passages with a marker. Those methods are easier and that is probably why they are popular but the time-pressed student needs a better way. Practice testing is that way.

The practice test gives you a baseline of your learned knowledge and identifies areas that need further study. Missing a question and discovering the correct answer provides an engagement with your memory that creates new neural pathways. "We learn from our mistakes" is not just an old saying. It is the truth.

The biological architecture of your brain suggests that memory functions like a muscle that grows stronger through the struggle of trying to remember something difficult (Bjork & Bjork, 2011). When you get a question wrong on a practice test, it creates a unique mental gap that makes you pay much more attention to the correct answer once you see it (Kornell et al., 2009). This specific phenomenon, known by scientists as the **Testing Effect**, fundamentally works better for long-term learning than just reading your notes over and over again (Roediger & Karpicke, 2006).

One fascinating discovery is that trying to answer a question before you even learn the lesson actually helps you remember the information better later on (Kornell et al., 2009; Little & Bjork, 2016). This works because the act of searching your brain for an answer you do not have yet makes the real facts "stick" more easily when you finally find them (Kornell et al., 2009). Making an initial mistake prepares your mind to prioritize the right data (Little & Bjork, 2016).

The exact moment you realize you made an error serves as a high-energy event that tells your brain to save that specific correction (Metcalfe, 2017). Because correcting a mistake requires your mind to rebuild its understanding from scratch, the memory becomes much more durable than if the answer had been handed to you without any effort (Metcalfe, 2017). Providing yourself with immediate feedback after a failed attempt ensures that the wrong path is closed off while the right one is strengthened (Metcalfe, 2017).

Strangely, the errors you feel most "certain" about are actually the ones you are most likely to remember correctly in the future after you fix them (Butterfield & Metcalfe, 2001). This is called the **Hypercorrection Effect**, and it happens because the psychological shock of being totally wrong makes you focus intensely on the incoming truth (Butterfield & Metcalfe, 2001). Learn from every mistake. These high-confidence errors are usually linked to strong ideas, providing a better frame for the new, correct information to attach to (Butterfield & Metcalfe, 2001).

Certifiable Optician, THE BIG BOOK provides the reinforcement that makes the topics addressed in the ABO exam a permanent part of your memory and experience. That does not mean this is easy. The serious student shall work through the concepts methodically. Studies have also shown the "spaced repetition" concept for learning is highly effective. Twenty-five to forty-five minute sessions, with breaks between, will provide the most effective retention process.

Certifiable Optician does not pretend to cover every possible question on the ABO but it does cover the vast majority of possibilities. Most importantly, the Certifiable Optician book covers these topics in a way that helps visualize and reinforce the topics when used together. This is the most efficient pathway that **places great value on your time.** No fluff, just facts. Add in your real-life experience in optical and you should be well-positioned to pass your test the first time.

Introduction

THE greatest difficulty I have witnessed students having is **visualizing the concepts** that lead to certification. The effect of prism. The meaning of the optical cross. Finding the power on the 180. There is a leap to be made from the concepts that are lying cold on a piece of paper to understanding it in form and principle and application.

I have endeavored to turn the more difficult concepts (according to my students) into drawings and diagrams to aid in visualizing the concepts. The addition of visual context to an otherwise dry formula or concept makes understanding easier and more natural. Memorization is aided by the visual framework.

Since I handled my first lens blank in 1972 at a Bausch & Lomb lens factory in Rochester, NY to my MBA degree just a few years ago I have never been far from the optical business. I have sponsored apprentices and explained Prentices Rule countless times. There are two moments that bring the greatest satisfaction:

1. The child that looks through their first pair of glasses and looks around in joyous amazement at the new world.

2. The optical apprentice that senses the light bulb going on at the meaning of a tricky concept like minimum blank size or decentration to induce wanted prism.

Visualizing optical concepts reinforces understanding and builds a bridge to deducing an answer as opposed to relying on rote memorization. Indeed, memorization becomes easier with understanding. Working through the practice exercises builds the "muscle memory" you need to succeed.

It has been said that a picture is worth a thousand words. The Certifiable Optician takes that adage to heart and uses over 150 diagrams and illustrations to trigger understanding and visualization of core optician concepts.

This approach promises to add another layer to your understanding so when confronted with the concept in other studies or in a real-life situation you can draw on that understanding to narrow the choices you will have to make. That provides better results without painful repetition.

If you are serious about achieving certification you now have a valuable tool waiting. Visualizing optical concepts can begin today if you do not procrastinate. Start your journey to certification. You too can be a Certifiable Optician!

A Certifiable Optician

OPTICIANS bend light. And they get paid to do it. How much you get paid varies considerably depending on location. Why? Twenty-one U.S. states require some form of licensure to practice as an optician. In these states an optician, on average, can expect to earn several more dollars per hour more than in an unlicensed state (Effects of Occupational Licensing, 2015). Even this number can fluctuate considerably depending on the rigor involved as some licensed states are more difficult than others.

U. S. states with licensure are Alaska, Arizona, Arkansas, California, Connecticut, Florida, Georgia, Hawaii, Kentucky, Massachusetts, Nevada, New Jersey, New York, North Carolina, Ohio, Rhode Island, South Carolina, Tennessee, Vermont, Virginia and Washington. The balance of U. S. states does not regulate the training of opticians, but certification has value as it provides recognition of a fundamental skill set useful in the profession (Texas requires certification but not licensure).

The recognized certifications in the U. S. are the National Opticianry Competency Examination (NOCE) and the Contact Lens Registry Examination (CLRE). These certifications are also a portion of the pathway to licensure in some states that require licensure. They are commonly referred to as the ABO exams as in American Board of Opticianry.

A plethora of information on these certifications is an interwebs search away. Much of this information is useful and relevant but difficult to absorb without a concise guide. My mission is to cut through the clutter and present the **specific data needed to achieve certification** with a **laser focus on the relevant facts**, and nothing but the facts. Time is money and I will strive to minimize the time, and by extension, the money and effort you need to dedicate to achieving certification.

How to Use this Book

I recommend starting by skimming this book for concepts and jargon at least a couple times. Organize the concepts you need or are interested in and then dig into the details. If it is in this book you will need to know it. Review the following research on the science of learning to customize your approach. Hundreds of practice questions are included in this book to assure your success.

The Cognitive Mechanisms of Retrieval Practice

In cognitive psychology, the use of testing as a learning tool is known as the Testing Effect or Retrieval Practice. While often viewed merely as a method of assessment, research demonstrates that the act of testing itself is a potent modifier of memory, often yielding greater long-term retention than repeated study (Roediger & Karpicke, 2006).

1. The Shift from Passive to Active Processing

The most common study method—and often the least effective—is passive review, such as re-reading notes or highlighting text. This approach often leads to an "illusion of competence," where familiarity with the material is mistaken for mastery (Dunlosky et al., 2013).

In contrast, practice testing forces active retrieval. When a learner is required to answer a question, they must search long-term memory to reconstruct the information. This effortful retrieval process strengthens the neural pathways associated with that concept more effectively than additional exposure to the material (Roediger & Karpicke, 2006).

2. Desirable Difficulty

The concept of "Desirable Difficulty" suggests that learning tasks that require considerable mental effort—though they may slow down initial acquisition—lead to better long-term retention (Bjork, 1994).

Practice testing introduces this beneficial struggle. When a learner strains to recall an answer, the difficulty signals the brain that the information is salient. Successfully retrieving information after a delay interrupts the forgetting curve, reducing the rate of memory decay (Bjork & Bjork, 2011).

3. Metacognitive Calibration

Practice testing serves a critical metacognitive function by providing objective feedback on what a learner actually knows versus what they believe they know. Without testing, students often struggle to accurately judge their own learning, leading to inefficient study habits (Dunlosky et al., 2013).

Testing acts as a diagnostic tool, revealing specific knowledge gaps. This allows learners to allocate their study time more efficiently toward weaker areas rather than reviewing material that has already been mastered.

4. Organization of Knowledge

Contrary to the belief that testing only improves rote memorization, evidence suggests it also enhances conceptual learning. The anticipation of a test encourages learners to organize information more coherently. Research indicates that retrieval practice can help students create mental models that are just as effective, if not more so, than those created through elaborative study methods like concept mapping (Karpicke & Blunt, 2011)

Chapter One

Critical Measurements

THE first step for any optician is to become familiar with the portion of the metric system used for measurement of distance. Certification requires an understanding of linear measures from kilometers to nanometers. Once certified you will use the millimeter measurement every day. The other units will form context for some of the simple math needed and for an understanding of light and accurate lens delivery to patients. Given that **bending light is our number one task** it is critical to have a working knowledge of the measurements used in our profession.

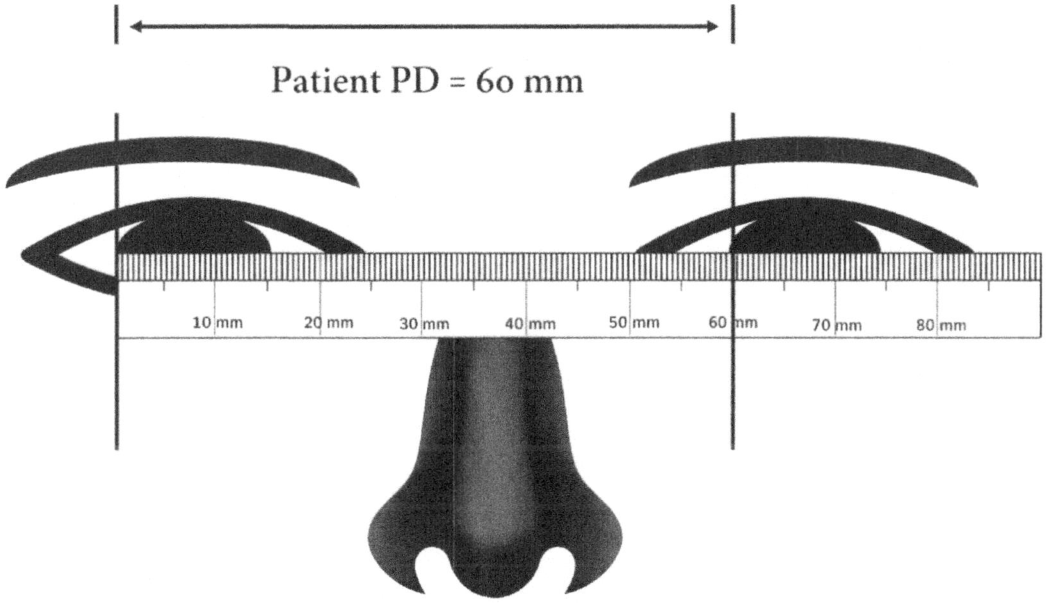

As you will come to understand, the most critical measurement an optician makes, day to day, is called the pupillary distance (PD). This is the distance between the centers of our eyes. This measurement can be made several different ways (PD Stick, Pupilometer, iPad etc.) but will always be recorded in millimeters. There are about 25 millimeters in an inch (25.4) but you will never specify this measurement in inches. 99 percent of PD measurements shall be between 50 to 75 millimeters (Dodgson, 2004).

Critical Conversions

Let us look at the metric system relevant to the optician. There is no need to memorize this but understanding the relationship is helpful. The meter is the base unit of length measurements in the metric system. All other units are in factors of ten. Visualizing these units is the first step to understanding. A kilometer is large. A meter is a tad longer than a yard. A millimeter is about 1/25th of an inch. A nanometer is one billionth of a meter. Very small!

Metric	Approximate English Equivalate
One Kilometer	39,370 inches
One Meter	39.4 inches
One Millimeter	4/100^{ths} of an inch
One Nanometer	.000000004 of an inch

Metric	Relative to the meter
One Kilometer	1,000 meters
One Meter	One Meter
One Centimeter	1/100th of a meter
One Millimeter	1/1000th of a meter
One Nanometer	1/1,000,000,000th of a meter

Understanding the very basic math involved in converting these units for use in optician formulas will serve you well as you move towards certification, but it starts with understanding the following.

100 Centimeters (cm) = 1 Meter (m)

10 Millimeters (mm) = 1cm

No short cuts here. Do what you need to be familiar with this relationship.

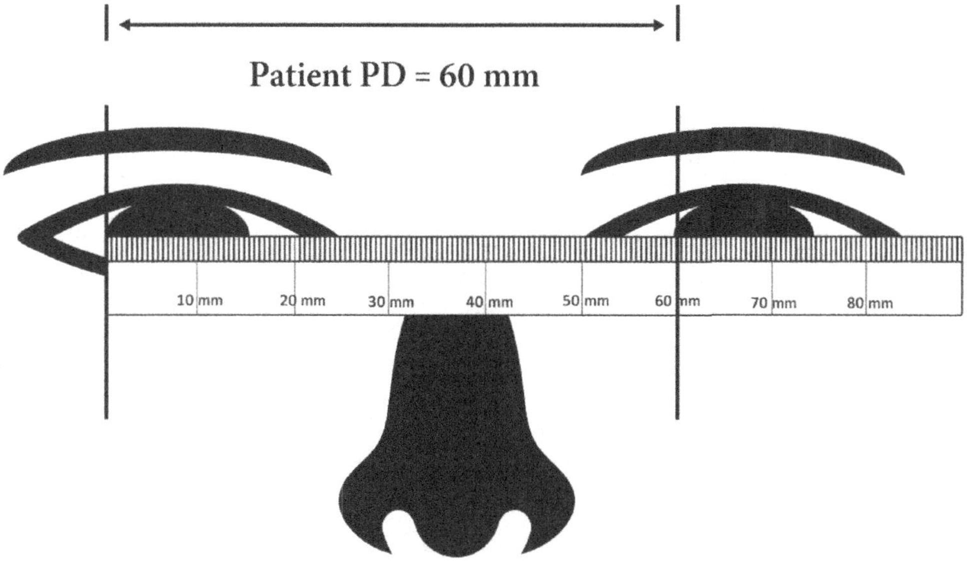

Take this measurement with a common PD Ruler. The measurement is 60 millimeters. What would this be in centimeters or meters?

mm	cm	m
60	6	.06

60 mm/10 = 6 cm

6cm /100 = .06 m

60 mm /1000 = .06 m

Notice that as you convert units from mm to cm to m (**going larger**) you divide by the relationships you just memorized.

m	cm	mm
.06	6	60

- .06 m X 1000 = 60 mm
- .06 m X 100 = 6 cm
- 6 cm X 10 = 60 mm

Notice than as you convert units from m to cm to mm (**going smaller**) you multiple by the relationships you just memorized.

Visualize the ruler in the picture. The PD is 60 mm OR 6 cm OR .06 m. This is 90 percent of the metric measurement concepts you will need to become certified. 75 percent will get you a passing score on the questions that involve metric conversions. Try the conversions with other numbers like 56 mm or 72 mm.

Soon you will see an additional (very, very small) metric measurement called nanometers (nm). A nanometer is one billionth of a meter. You will never use this as a day to day optician, but it will help in your understanding of light. Remember, opticians bend light for a living, so it is essential to understand light and how to measure it.

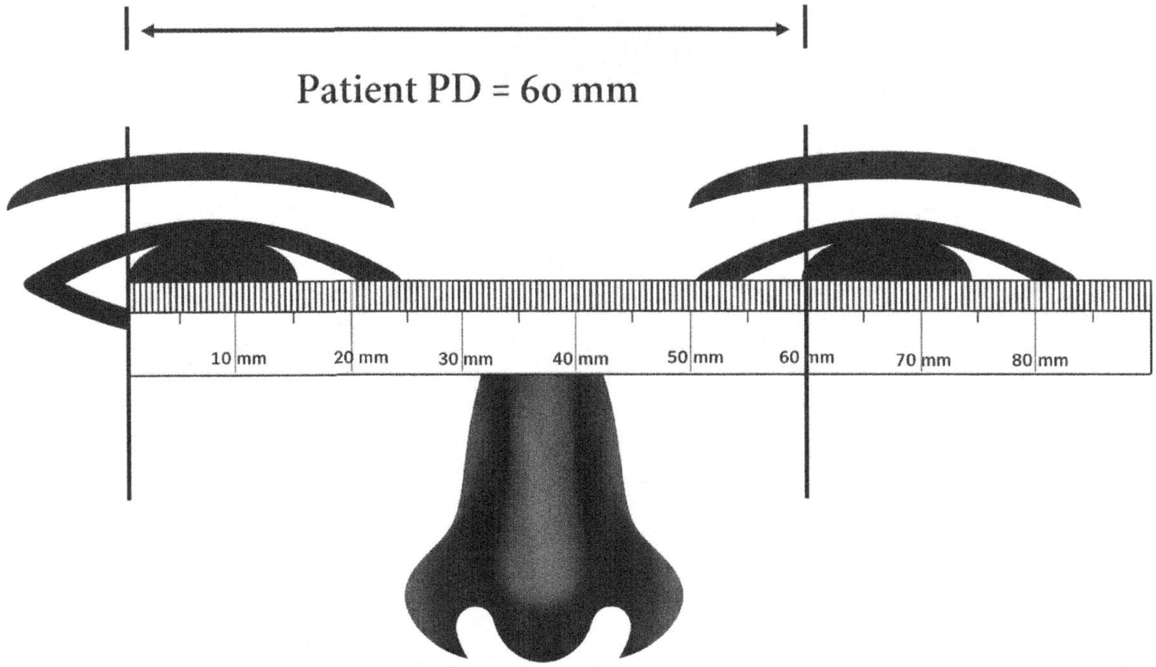

Chapter Two

Opticians Bend Light

What is light?

WIKIPEDIA has as good a definition as needed for opticians; "…visible light is electromagnetic radiation within the portion of the electromagnetic spectrum that can be perceived by the human eye" (Light, n.d.).

Light in a vacuum will travel at 186,000 miles per second or nearly 300,000 kilometers per second. As soon as light enters a clear medium like glass or water it slows down and refracts (bends). It is this slowing and refracting that gives an optician the power and ability to fill an Rx. You will get comfortable using this knowledge to bend light and improve your patients' vision.

You have probably also learned that white light is composed of all colors. You will need to understand the visual spectrum and the wavelengths of each color that makes up white light.

Wavelength is the distance between the crests of light waves, just like the ripples in a pond when you throw a stone. The wavelength of light is measured in nanometers (nm); the unit is one billionth of a meter.

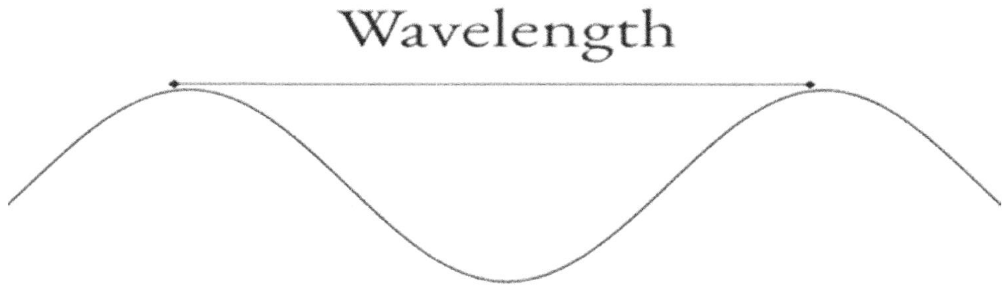

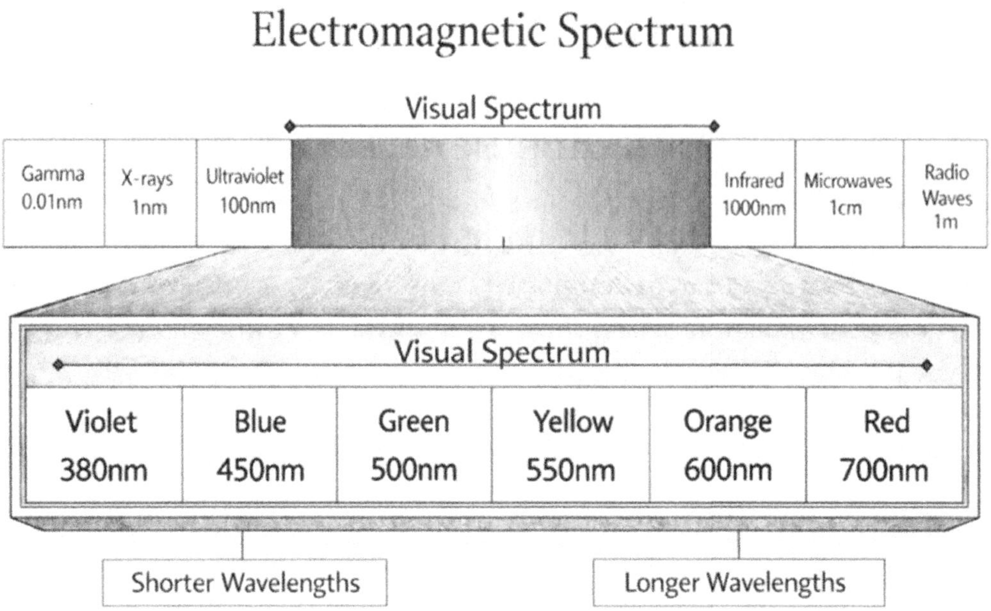

Source: Helmenstine, A. M. (2020, Septtember 10). *Understand the visible spectrum of light.* Retrieved from https://www.thoughtco.com/understand-the-visible-spectrum-608329.

↑ Color the VISUAL SPECTRUM Boxes ↑

This is how the different colors fit into the overall electromagnetic spectrum. In this chart, to the left are the high energy shorter wavelengths like gamma rays and ultraviolet. To the right are the longer wavelengths up to the long, lower energy, like radio waves.

Take note that the visual light our eyes detect is from about 380 to 700 nm. Shorter wavelengths like ultra-violet (UV) have been proven harmful to the eye and most eyecare products take pains to block or reduce it. UV is what tans your skin. It is best left out of the eye. More recent research has demonstrated that even the blue light from computer screens and cell phone may cause negative effects. Blue light blockers are available in most optical locations now.

Note the color in the center of the visible light spectrum, yellow. That center color is the easiest for the eye to detect, ergo, school bus yellow. The opposite end from blue light is the red longer wavelengths that moves into infrared. Ever see those red lamps over the fast food french fries? Infrared lamps put out the heat that keeps those fries crispy and warm. Night vision goggles rely on detecting infrared or heat from the body.

Concentrate your learning on understanding that blue light is the shorter, more energetic wavelength, yellow is dead center at about 560nm and the red waves get longer. This understanding will be important when we discuss color breakup in a prism. Next time you see a rainbow understand that the light rays are bent by the water vapor in the air and that separates (refracts) those colors into the beauty of the rainbow. The **color separation is dependent on the wavelength and energy** of the different colors.

The Magic in the Mist: How Science Paints the Sky

Have you ever wondered how a grey storm turns into a vibrant arc of color? The secret lies in the interaction between sunlight and water. Every rainbow begins with a single raindrop acting as a tiny liquid prism, bending and reflecting light to paint the sky with a brilliant spectrum.

A primary rainbow forms when sunlight enters the top half of a raindrop. As it enters, the light undergoes refraction, slowing down and bending. Because "white" light is a mixture of wavelengths, it undergoes dispersion: violet light has a shorter wavelength and bends the most, while red light has a longer wavelength and bends the least.

After a single internal reflection off the back of the drop, the light refracts again as it exits the front. Red light leaves the drop at a steeper angle of approximately 42° relative to your line of sight, while violet exits at about 40°. Because of this geometry, you see red light coming from higher drops in the sky and violet from lower drops, placing red on the outer edge of the arc (National Institute of Standards and Technology [NIST], 2024).

Double Rainbows: The Power of the Second Bounce

Occasionally, sunlight enters the bottom half of a droplet, creating a more complex path. In a double rainbow, the light undergoes two internal reflections instead of one. This "second bounce" triggers a dramatic reversal: because every reflection flips the color order, the secondary arc appears mirrored, with violet on the outside and red on the inner edge. Furthermore, because some light "leaks" out the back of the drop during every bounce, the extra reflection causes significant energy loss. This makes the secondary arc appear much fainter than the primary bow.

Alexander's Dark Band: The Gap in the Glow

Between these two arcs lies a mysterious shadow known as Alexander's Dark Band. This region appears noticeably darker than the rest of the sky because the light that would normally occupy that space is being "refracted away" into the two separate rainbow arcs. This leaves a "gap" where very little light is reflected back to your eyes, creating a deep contrast that defines the two glowing arcs (Thompson & Taylor, 2008).

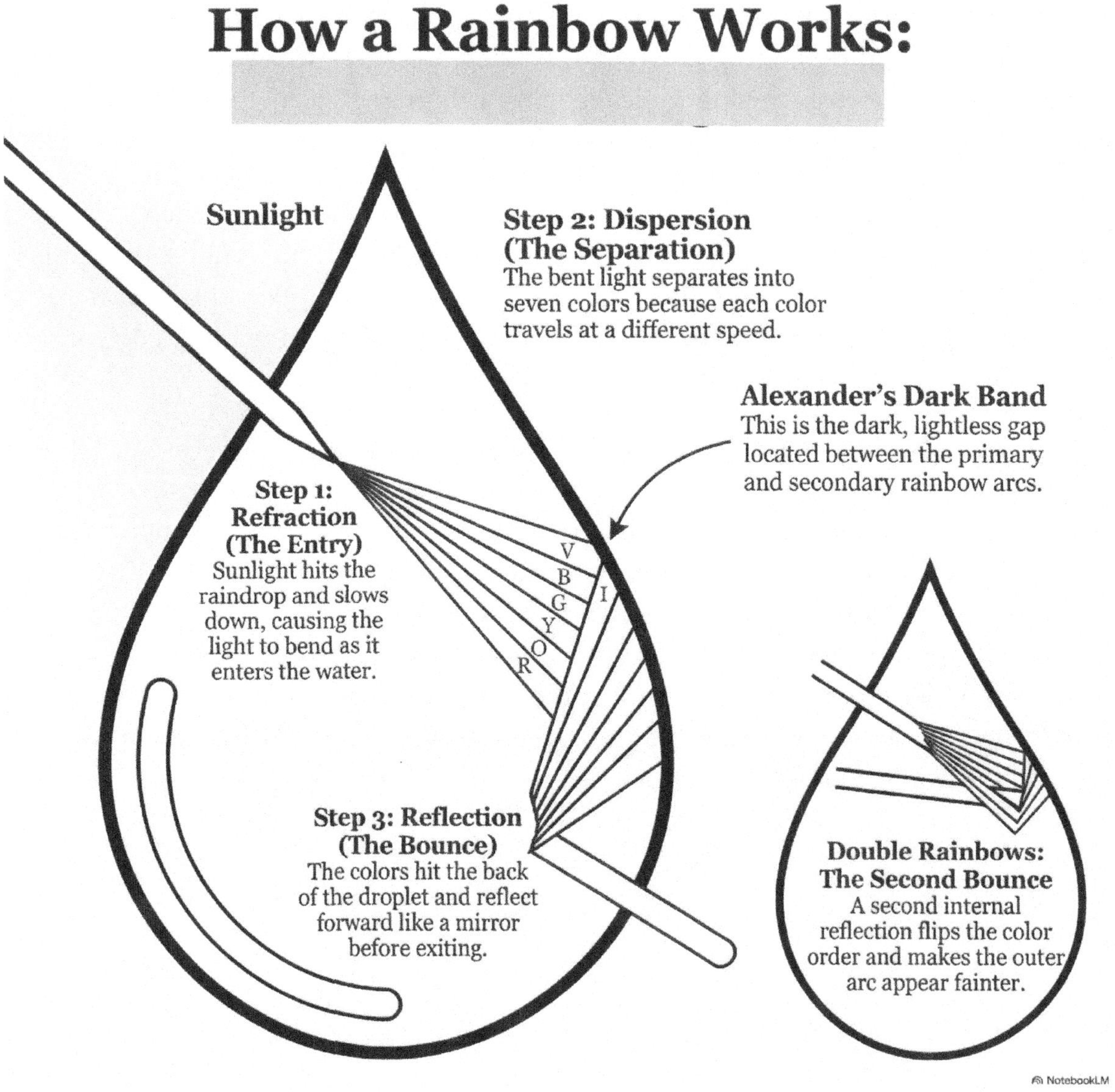

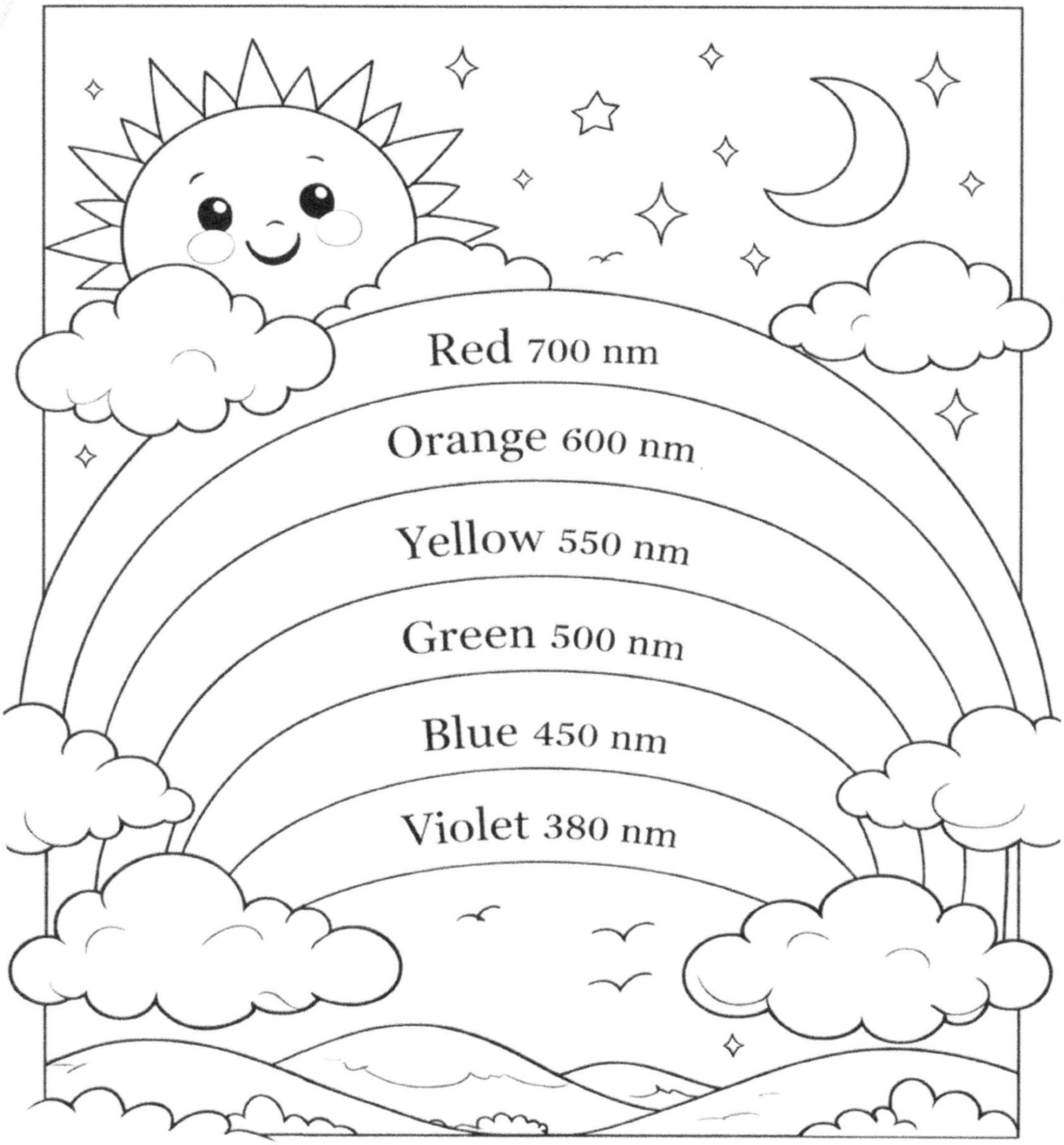

Color this image.

Rules of the Rays

Diagrams of the path of light illustrate how opticians bend light to achieve the goal of helping people see better. Ray tracing can help you visualize the function of the various lenses and/or prisms that will become part of your everyday work life.

Although light travels in all directions unless directed (like a laser) we graph the behavior of light in lenses as if it is a single ray or group of rays (beam). When ray tracing light, rays travel from the left to the right with direction indicated by an arrow.

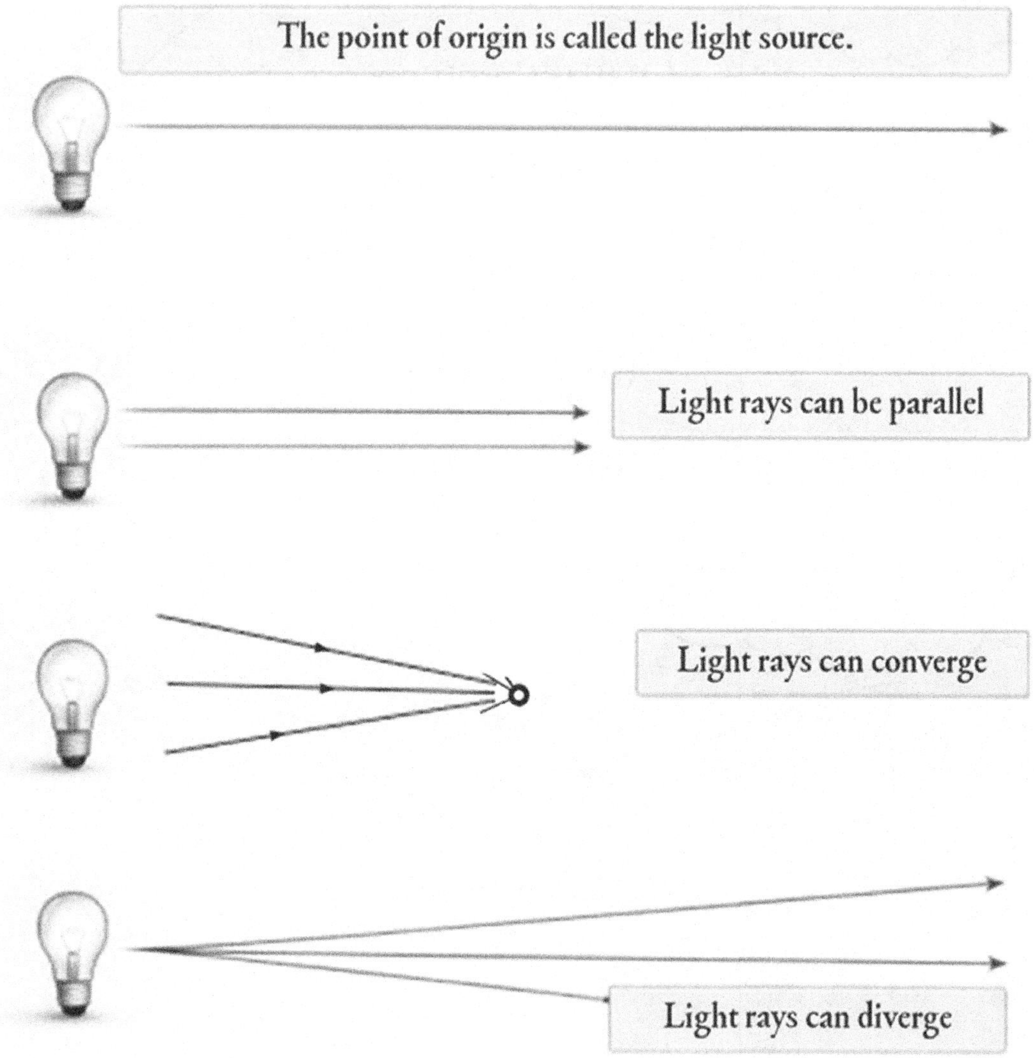

Chapter Three

Mirrors

THE optician manipulates light every day, generally, to converge or diverge light to accurately fill the eye doctor's prescription. That is the prime function of eyeglass lenses or contact lenses. Before we start discussing how a clear lens redirects light, we need to understand the effect a mirror has on a light ray.

The Flat or Plane Mirror

The key takeaway from a plane mirror is to understand that every light ray is reflected at an angle equal to its approach angle. The concept of a mirror image is that the reflection is virtually identical but reversed. Write your name on a piece of paper and hold it in front of a mirror. The reflection still shows your name; written backwards. The approaching light ray strikes the mirror at an angle called the Angle of Incidence. It returns to your eye at an Angle of Reflection. As the diagram indicates, the angle of incidence equals the angle of reflection.

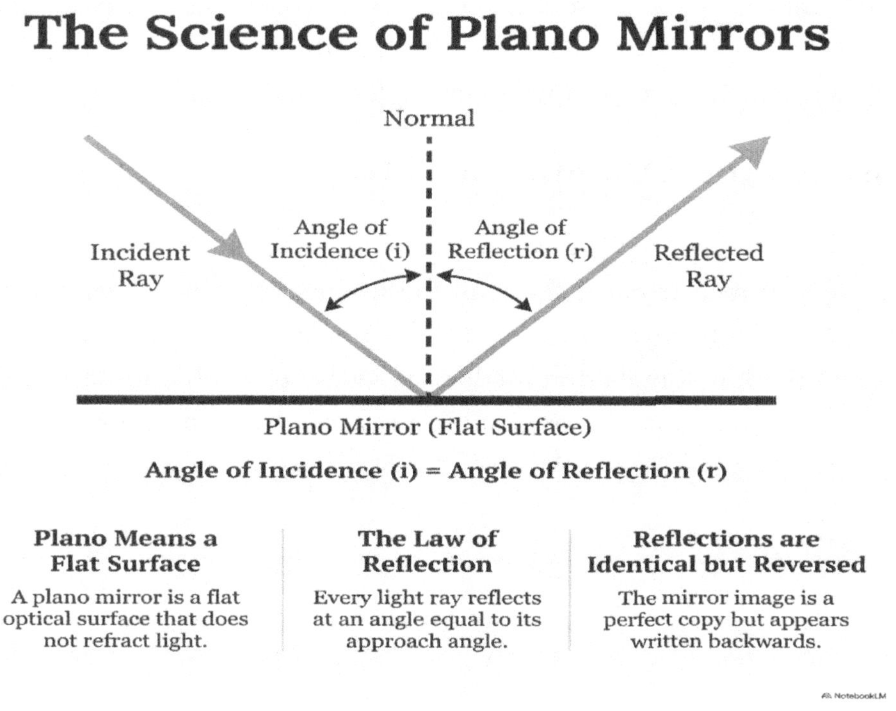

Why Mirrors Don't Actually Flip Left-to-Right: The Physics of Plano Surfaces

A "plano" mirror refers to a perfectly flat optical surface. While the term defines lenses with zero refractive power, in reflection, it signifies a surface that obeys the fundamental Law of Reflection: the angle of incidence equals the angle of reflection ("Principles of Plane Reflection and Plano Surfaces," n.d.). This simple geometry creates a persistent cognitive illusion regarding horizontal reversal that challenges our spatial intuition.

The Secret: Front-to-Back, Not Side-to-Side

Physics reveals that a flat mirror does not actually flip images horizontally or vertically; instead, it reverses the Z-axis, or the dimension of depth (Hewitt, 2014).

* The Forward Point: If you point your finger "North" toward the glass, the finger in the reflection points "South" directly back at you.

* Vertical/Horizontal Stability: Orientation remains stable; waving your right hand moves the hand on the right side of the mirror, and nodding up remains an upward motion.

The perceived "reversal" occurs only because observers mentally project themselves 180 degrees into the mirror space by rotating around a vertical axis (Serway & Jewett, 2018). Because the image is only inverted front-to-back without actual rotation, your right hand appears exactly where a "turned around" person's left hand would be.

Perception and the "Ambulance" Effect

The brain creates a "virtual image" by extending light rays behind the glass. Because the surface is plano, these rays remain parallel, allowing the brain to perceive a straight-line projection (Walker, 2014). This depth reversal causes the "Ambulance" effect. Consider the letter "L": the mirror reflects the horizontal arm exactly as positioned, but because the image is reflected toward you, the arm points in the opposite direction relative to your perspective (Walker, 2014).

Conclusion : Brain Logic vs. Physics

While the physics of a plano surface only involves a reflection of depth, our brain's interpretation of that data creates the illusion of a horizontal flip. Understanding this distinction clarifies why mirrors seem to defy logic while strictly following the laws of reflection.

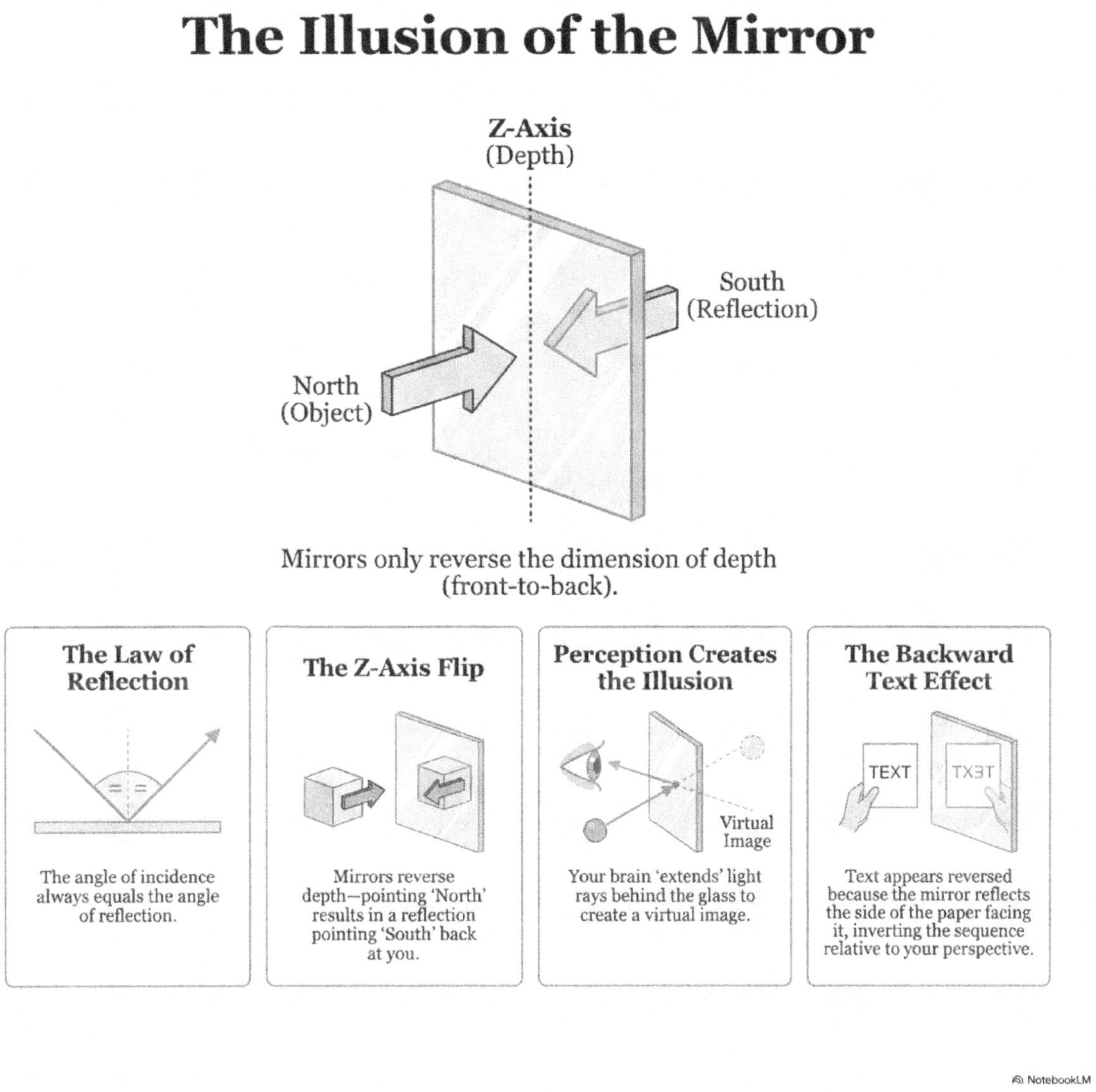

The Convex Mirror

A curved surface, be it a mirror or a lens, is where we can begin to bend light in a way that serves a more complicated purpose than a flat mirror. We have all seen the convex mirror used as a security feature in stores or on our vehicle's side mirror (objects may appear farther away).

The amount of curvature will dictate the size of the reflected image and field of view. The image on a convex mirror will be smaller and that is also why the field of view is larger. The smaller image creates the illusion that the image is behind the mirror. In physics this is referred to as a virtual image. When you see dotted lines like the example, it indicates a path to a virtual image, the distance to the virtual image can be used in the calculations of curvature and power.

A convex mirror, characterized by its outward-curving reflecting surface, functions as a diverging mirror (Hewitt, 2014). When parallel light rays strike this surface, they reflect outward and away from a central axis, never physically intersecting in real space (Serway & Jewett, 2018).

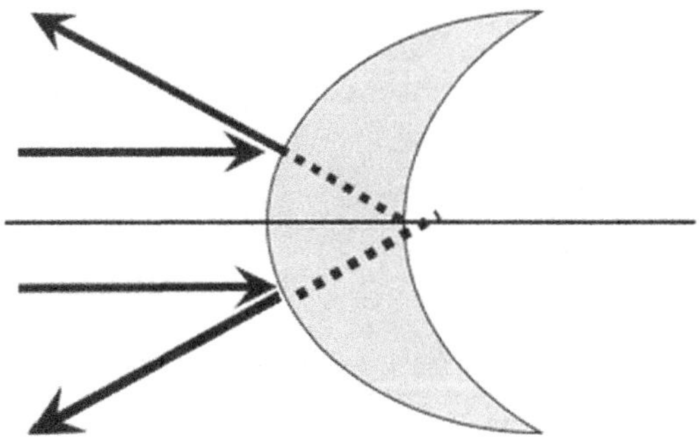

Because these rays diverge, the human brain traces them backward to an imaginary intersection point behind the mirror's surface known as the virtual focal point (Walker, 2014). This geometric process results in an image that is always virtual, upright, and diminished (smaller) than the actual object. The utility of this "shrinking" effect is that it compresses a significantly wider field of vision into a small area, which is why these mirrors are essential for vehicle side-view safety and store security (Serway & Jewett, 2018).

Objects appear smaller. The field expands. The "Ambulance effect" occurs because a convex mirror reverses the Z-axis (depth) while maintaining vertical and horizontal orientations (Hewitt, 2014). Consequently, letters facing the mirror appear flipped from the driver's perspective. To ensure legibility in a reflection, text is painted "backwards" on vehicle hoods, allowing the mirror's front-to-back inversion to restore the correct sequence (Walker, 2014).

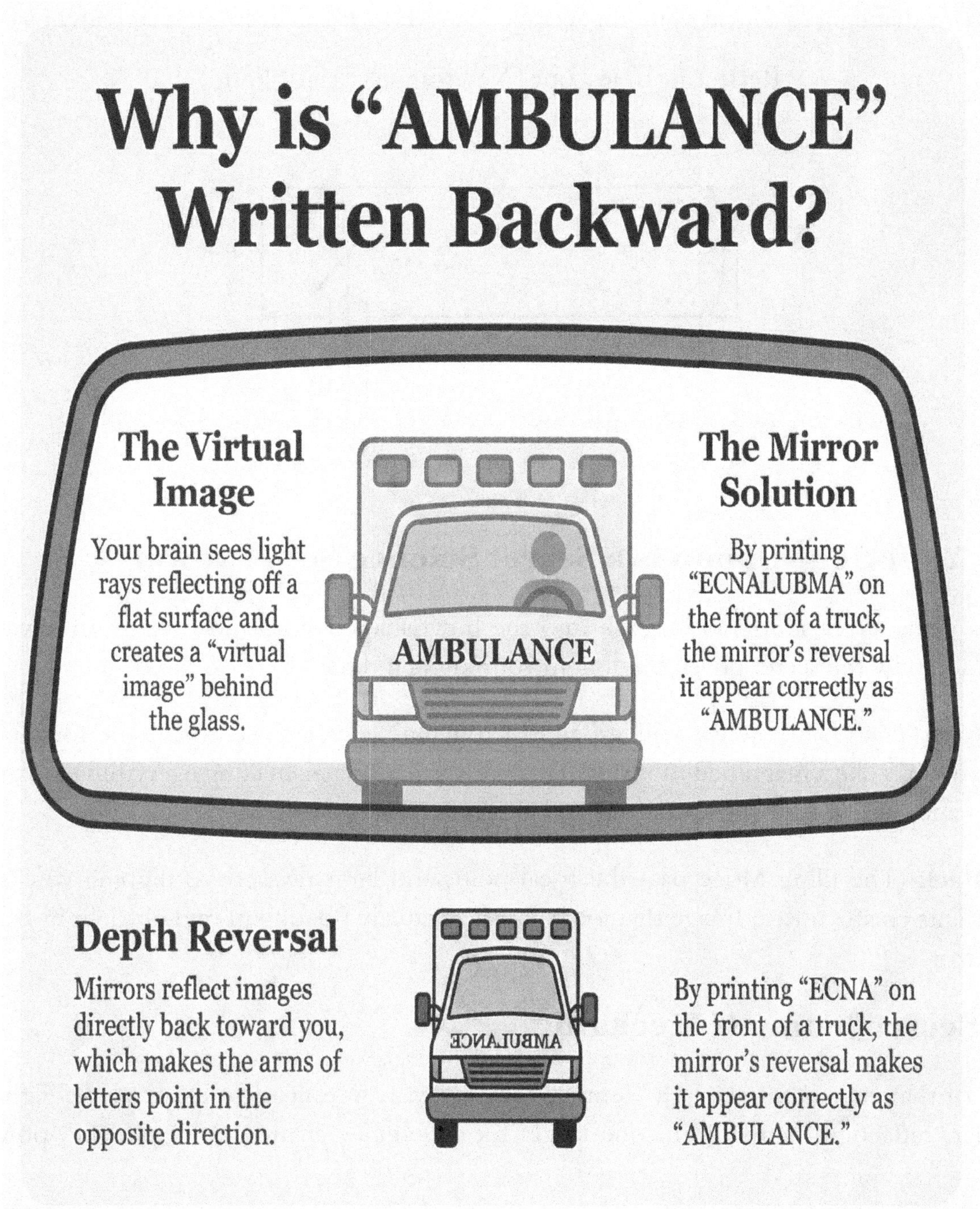

The Concave Mirror

A concave or converging mirror gathers light and reflects the image to a focus point in front of the mirror. It forms a "real image". Reflecting telescopes gather light from distant objects and directs that light to a focal point that can be observed with an eyepiece that contains lenses.

Reflecting Telescope (Newtonian): Light Path

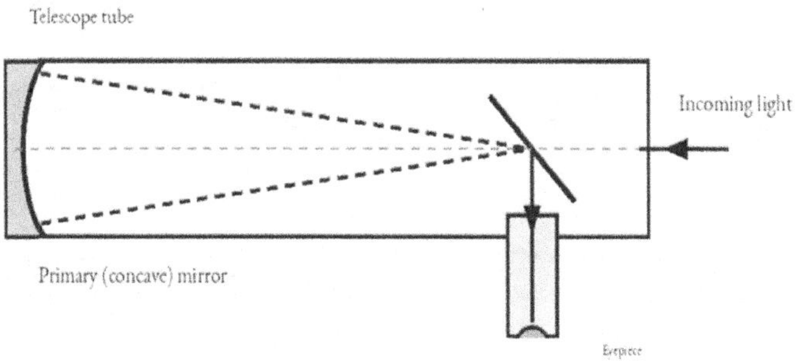

Light reflects from the primary mirror, converges toward the focus, and is redirected to the eyepiece by the secondary mirror.

Dashed rays indicate the converging beam after reflection from the primary mirror.

Why Your Soup Spoon is a Secret Science Superpower

Grab a shiny silver soup spoon, look into the bowl-shaped side, and prepare to have your mind blown by the secret physics hiding in your kitchen drawer!

Up Close (The Giant): When you are almost touching, super close (inside the focal point), your eye looks like a magnified monster! This is a Virtual Image, making everything look right-side up and huge; exactly how makeup and shaving mirrors work.

Step Back (The Flip): Move past the focal point and light rays cross, flipping you upside down! This creates a Real Image that looks like it's actually floating in mid-air right in front of the mirror.

The Secret "Funnel" Mechanic

Think of that curved surface as a "funnel" for energy. It wrestles scattered, parallel light rays together, reflecting them toward one single focal point to create a concentrated, powerful beam.

Flashlights: Focusing light into a strong, directed beam.

Telescopes: Gathering distant starlight so we can see the deep universe. Automobiles have long employed curved mirrors. Your passenger side mirror is typically convex to give the driver

a larger field of view. Your vehicle headlights use a concave mirror to focus the beam forward, reducing the scattering of the headlamp light rays into a focus point well ahead. Mirrors make driving safer

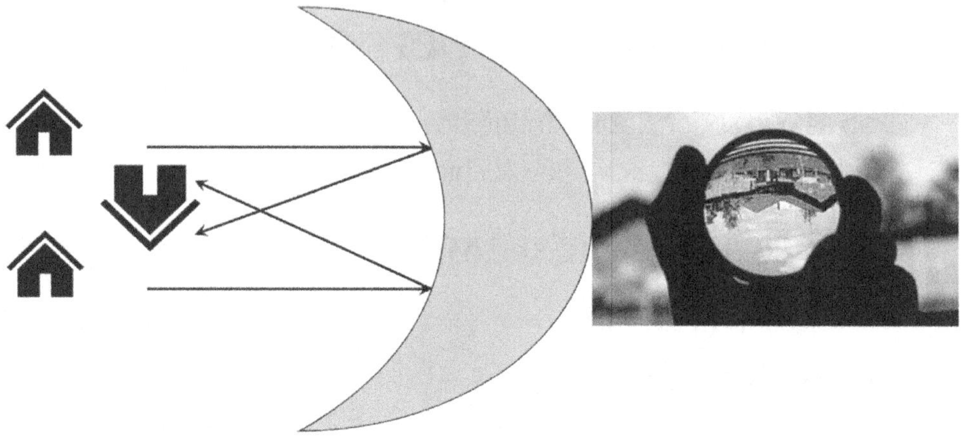

Think of a concave mirror as a "light bender" that forces rays to cross paths. Because the mirror curves inward like a bowl, it funnels all the light hitting it toward a single spot called the focal point. The inversion happens because of what the light does after it hits that point:

The Cross-Over: Light from the top of an object hits the mirror and reflects downward through the focal point.

The Swap: At the same time, light from the bottom hits the mirror and reflects upward through that same point.

The Flip: Once the light rays pass the focal point, they have officially swapped places. The light that started at the top is now at the bottom, and the light from the bottom is now at the top.

If you are standing far back, your eyes catch the light after it has made this switch, making you look upside down. However, if you move very close (inside the focal point), the light hasn't had the chance to cross over yet, which is why you appear right-side up and huge!

Distance changes the view.

Chapter Four

Prism

WHILE the mirror reflects light, any clear material refracts light. Bending light and refracting light are synonymous for our purposes. The prism shape is central to everything an optician does

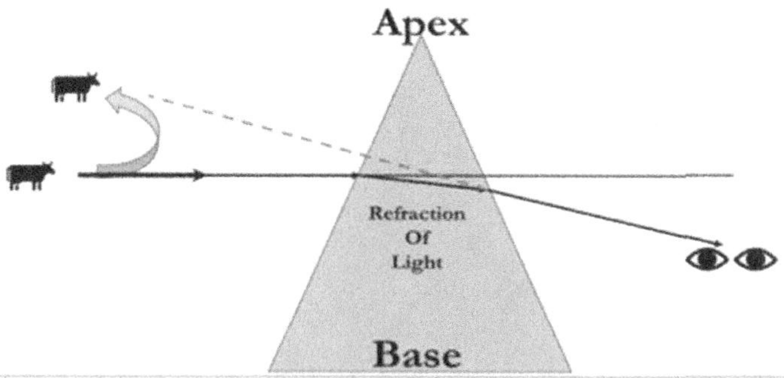

Light rays deviate (refract) towards the base when they enter the prism (slowing down more where thicker) and when they leave (speeding up as they exit the thinner portion). Because of this the object appears to move towards the apex.

The relationship between the light ray and the angle of the prism sides (and the material the prism is made of) dictates the amount of refraction. As the ray hits the angled surface it turns. Imagine walking into a cornfield, at an angle, while holding a broomstick parallel to your shoulders in front of you.. One end would hit the cornstalks first, turning you towards the cornfield. When the light ray leaves the prism one end of the broomstick hits air first and speeds up, turning you again.

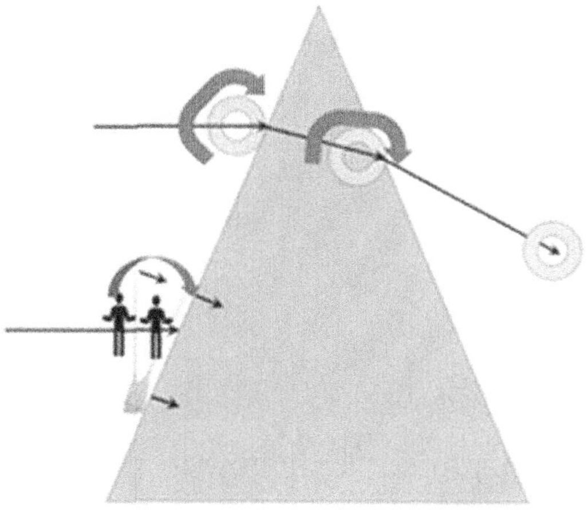

More than simple refraction occurs when light reaches a prism. Remember we discussed the different colors that make up white light. Each color has its own wavelength measured in nanometers. These differing wavelengths represent slightly differing energy levels and that in turn can result in different amounts of refraction. This element of refraction is called dispersion.

Dispersion

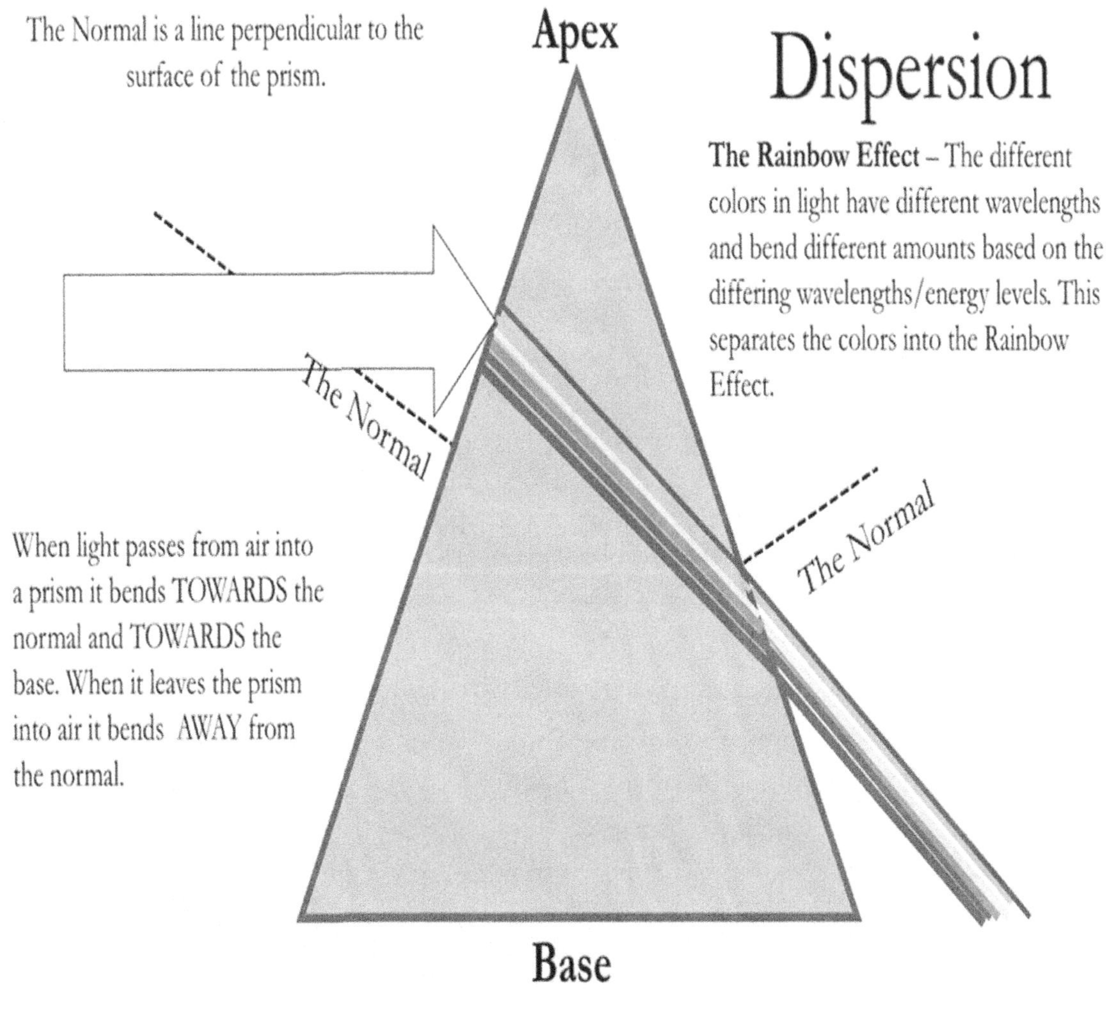

When you see a rainbow in the sky you know that light passing from air into water vapor (acting like a prism) and out again, is the combination of refraction and dispersion.

Armed with the knowledge of how light behaves in a prism you have the basics of how an eyeglass lens or contact lens correct vision. The devil is in the details.

Color the dispersion bands just like the RAINBOW

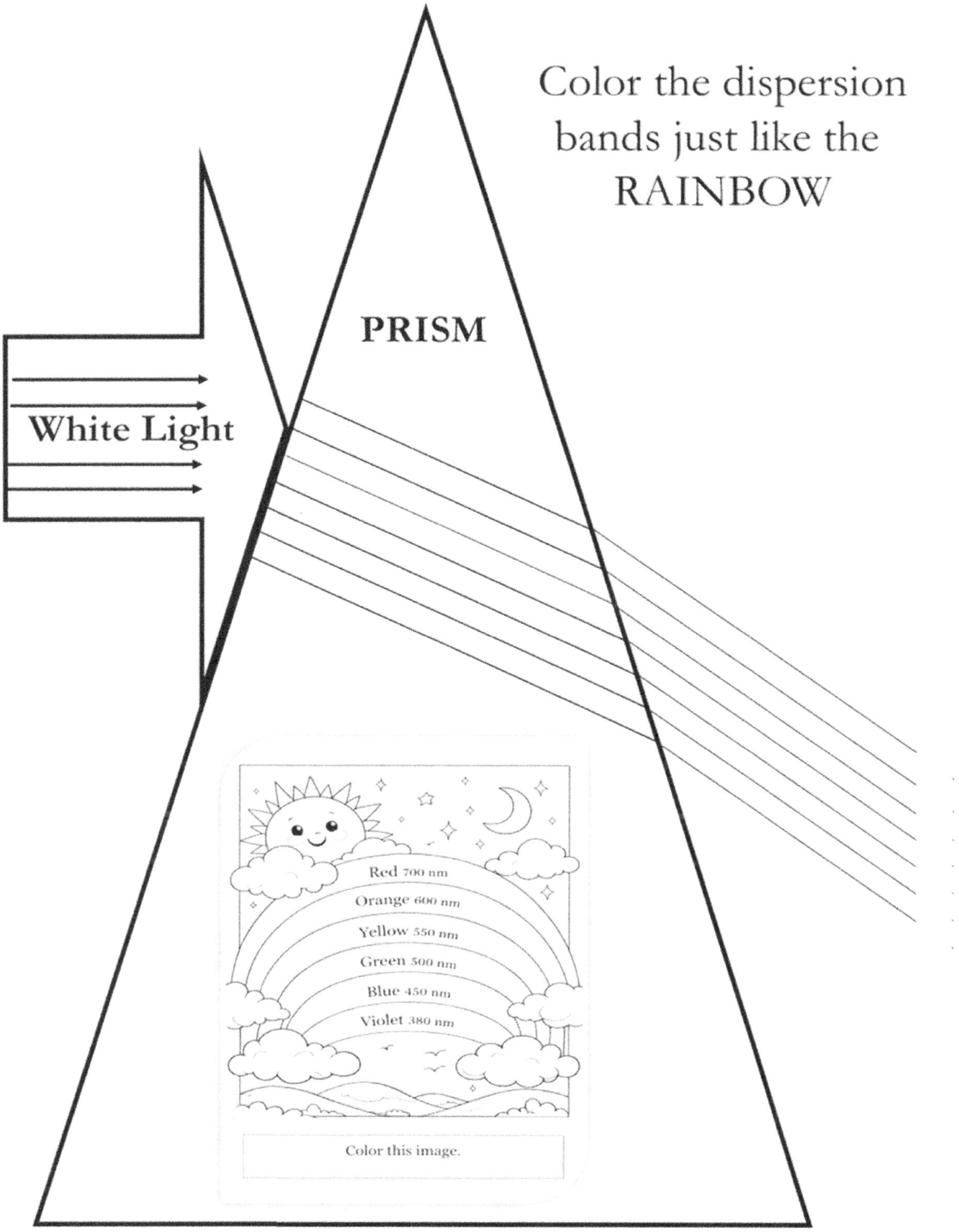

The Path of Light in a Prism

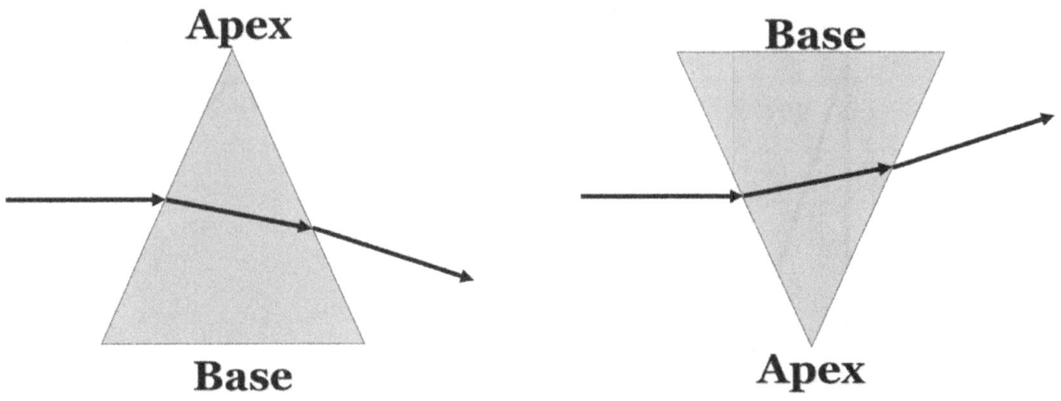

The devil is in the details; and how you put the prisms together!

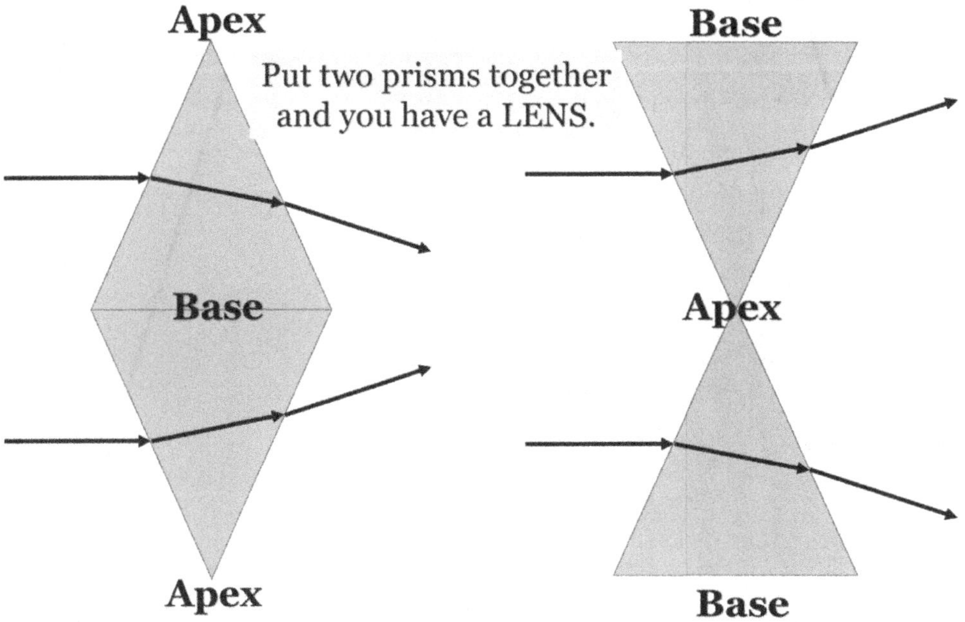

Snell's Law

Snell's Law is a mathematical basis for refraction within a prism. Within the "Ophthalmic Optics" studies, which constitutes approximately 25% of the modern ABO exam, the conceptual residue of Snell's Law is visible through questions regarding **refractive index** and **lens thickness** (ABO-NCLE, 2025). Candidates must understand that higher index materials (n) bend light more sharply, an observation dictated entirely by the underlying trigonometric relationship established by Snell (Brooks & Borish, 2007). However, **explicit math involving trigonometry is vanishingly rare in the current testing environment** according to recent candidate feedback and practice analysis (OptiBoard, 2024). Review the Snell's Law concept for general awareness as opposed to detailed application abilities.

A 1-diopter prism is very thin and only bends light a tiny amount, shifting the image just 1 centimeter for every 100 centimeters the light travels (Brooks & Borish, 2007). Because the wedge is so shallow, the light only hits the surface at a small angle, meaning it does not have to change direction very much to get through to the other side (Jalie, 2016).

In contrast, a 10-diopter prism is much thicker and has a very sharp, steep angle at its base (Brooks & Borish, 2007). This thicker wedge forces the light to hit the glass at a much steeper angle, which triggers a much larger "bend" according to Snell's Law (Jalie, 2016). Because this prism is ten times stronger than the first one, it shifts the image a full 10 centimeters over that same distance (Jalie, 2016). This creates a very noticeable jump in where things look like they are located when you peer through the glass (Brooks & Borish, 2007).

Snell's Law: The Primary Engine of Refraction

The core mathematical formula (Snell's Law) driving every prism calculation is expressed as:

$$n_1 \sin(\theta_1) = n_2 \sin(\theta_2)$$

Because the "bend" described by Snell's Law is what creates the displacement, we can translate these angles into the practical Prism Diopter Formula.

$$P = \frac{100 \times d}{f}$$

- P: The resulting power in prism diopters (Δ).
- d: The displacement of the light ray measured in centimeters.
- f: The distance from the prism to the target, typically 100 centimeters.

Example 1: The 1-Diopter Prism

In a 1-diopter prism, the angle of the wedge is very shallow, causing a minimal change in the light's path according to Snell's Law.

$$1\Delta = \frac{100 \times d}{100 \text{ cm}}$$

By solving for d:

$$d = \frac{1\Delta \times 100 \text{ cm}}{100}$$

$$d = 1 \text{ cm}$$

The light ray shifts exactly 1 centimeter away from its original path because the low-angle refraction is subtle.

Example 2: The 10-Diopter Prism

In a 10-diopter prism, the glass wedge is significantly thicker at the base, forcing the light to hit the surface at a much steeper angle. Snell's Law dictates that this steeper incident angle results in a much more aggressive bend.

$$10\Delta = \frac{100 \times d}{100 \text{ cm}}$$

By solving for d:

$$d = \frac{10\Delta \times 100 \text{ cm}}{100}$$

$$d = 10 \text{ cm}$$

The light ray shifts exactly 10 centimeters from its original path. Witness the geometric leap. This larger displacement is a direct result of Snell's Law acting on a steeper wedge, allowing opticians to correct major eye misalignments.

Snell's Law, cont.

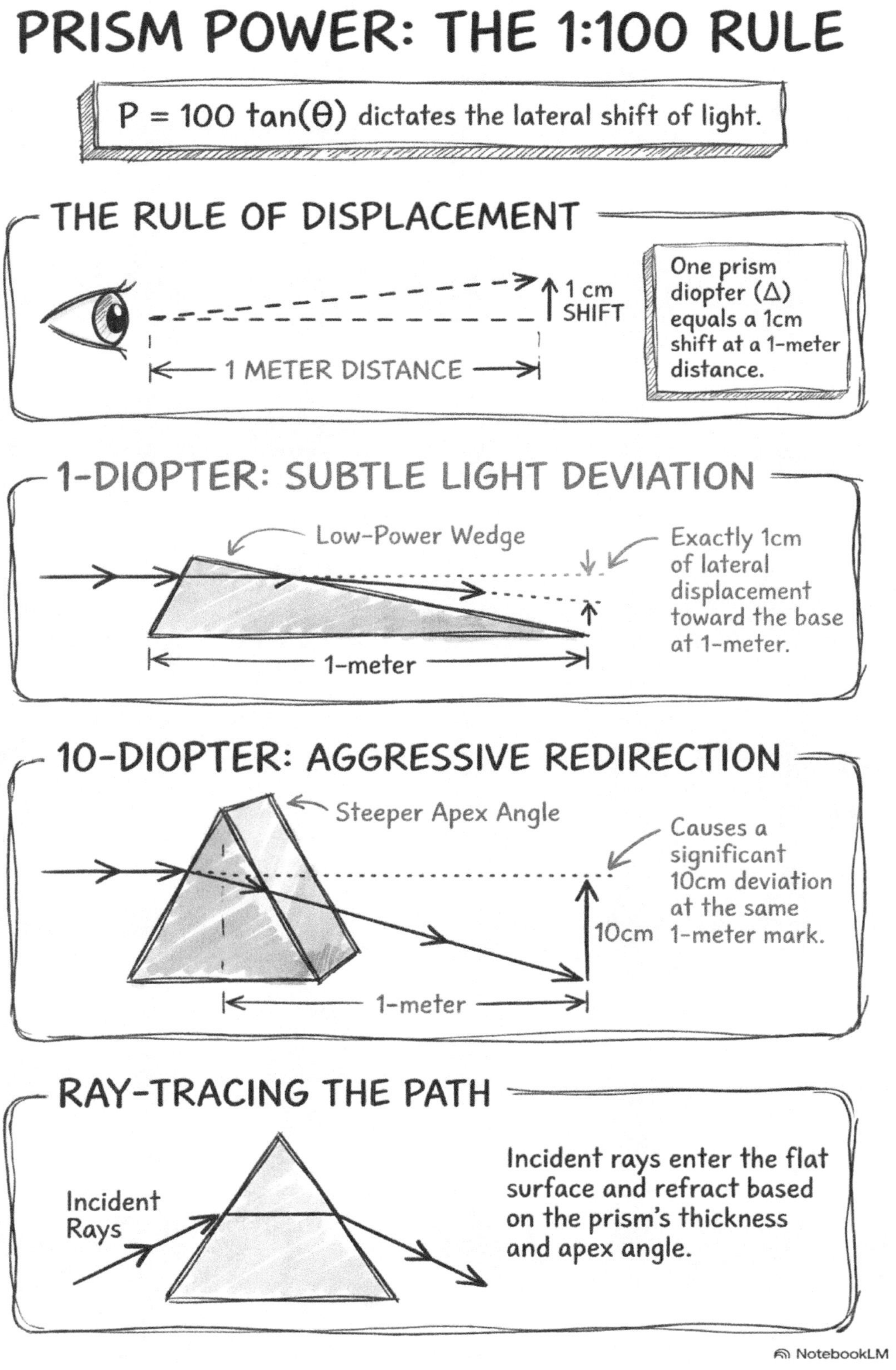

Chapter Five

Lenses are Two Prisms

LENSES are prisms set together. In tandem they become the lenses you see in eyeglasses. This is how opticians bend light to fill prescriptions.

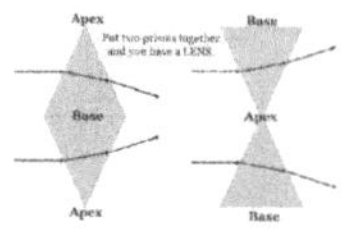

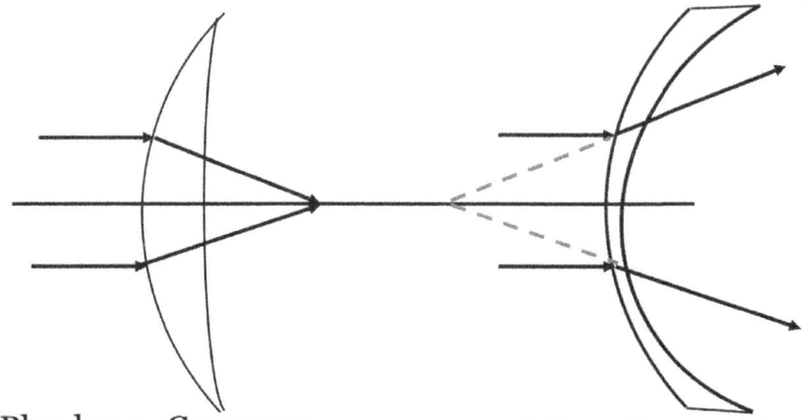

Plus lens - Converge   Minus lens - Diverge

These profiles of lens blanks, the form used to make eyeglasses, are refined versions of the prisms we discussed but the underlying theory is identical.

- **The Plus Lens (+)**
  - **Thicker in the center than the edges**
  - **Light rays converge to a real focus point**

- **The Minus Lens (-)**
  - **Thicker on the edges than the center**
  - **Light rays diverge and create a virtual or "imaginary" focus point**

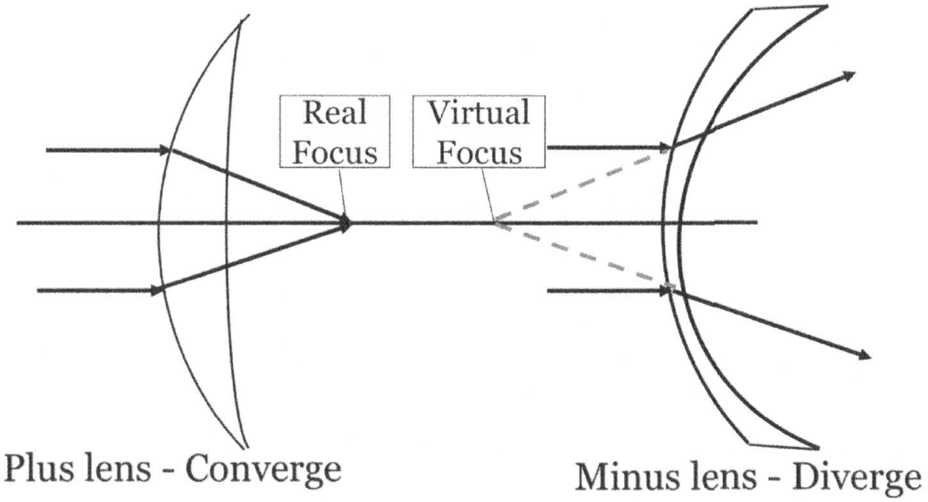

Note the dotted lines that track backwards from the diverging light rays on the minus lens. This is the concept of a virtual focus. Where is the point that the diverging rays would intersect if traced backwards? This distance is important in determining the power of lenses and allowing opticians to bend light to the benefit of their patients.

The geometry of a transparent medium fundamentally dictates how traveling light waves behave, similar to the way a glass prism splits a rainbow or a shaped mirror warps a reflection (Brooks & Borish, 2007). A plus lens bulges outward to become thicker in the center, forcing those incoming light rays to bend toward each other to help farsighted individuals focus on nearby objects. Conversely, a minus lens hollows inward to become thinner in the middle, causing the light rays to spread apart to help nearsighted individuals view distant scenery clearly (Jalie, 2016).

The Optical Center

There is one point on a lens where light does not refract. At this point light will neither converge nor diverge. That point is called the optical center (OC). Another way to look at it is the point where our two prims were put together, base to base or apex to apex.

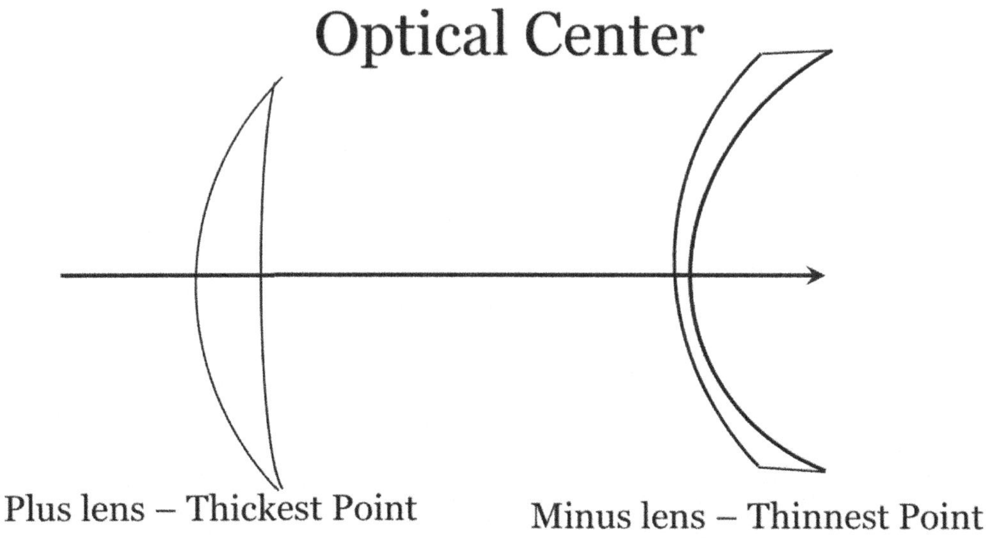

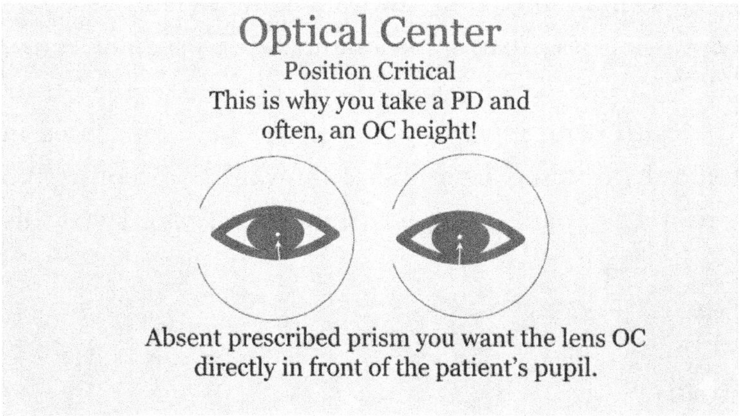

The optical center is the **thickest point** on a **plus lens** or the **thinnest point** on a **minus lens**. A light ray that enters the optical center, perpendicular to the front of the lens, shall not converge or diverge. It shall pass thru **undeviated**. Remember the PD we measured earlier? The OC's will need to be placed the distance equal to the PD. Otherwise the optician will induce unwanted prism and the glasses will fail inspection (unless the prism is prescribed).

Chapter Six

The Power of Lenses

The Diopter

THE amount that light will refract in a lens is dependent on the power of the lens and the material the lens is made of. This is how the optician bends light in a controlled fashion to accurately fill the doctor's prescription. Optician's measure the power of the lens in diopters. Diopters are a function of the focal length of the lens.

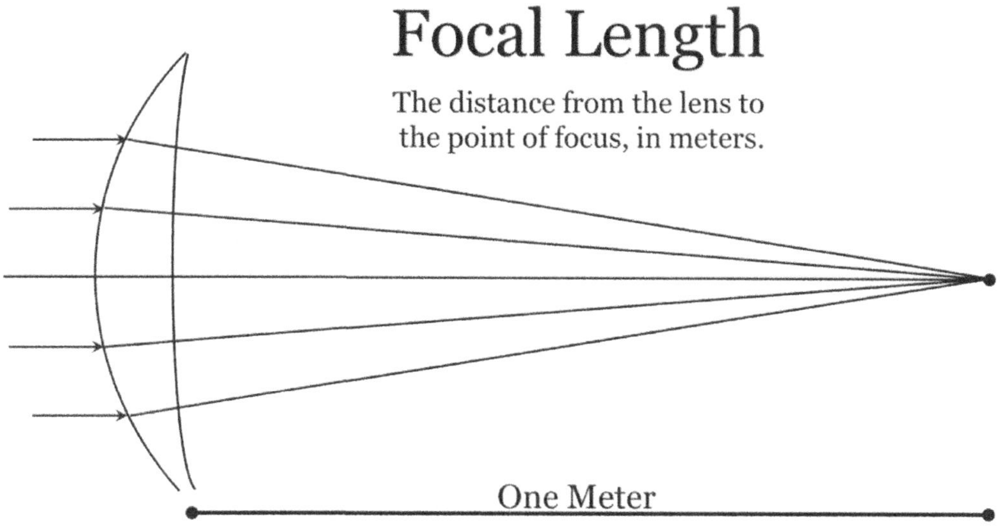

When you see a doctor's prescription for eyeglasses it will be written in diopters. A simple Rx might be written like this.

Right Eye	+1.00
Left Eye	+1.00

In your day-to-day life as an optician you will go to a drawer and take out a pair of +1.00 diopter lens blanks pre-made and ready to be glazed into a frame. But that will not suffice in your quest to be certifiable. There is math involved to understand this relationship.

The formula to determine the diopter power is **D=1/d** Plug in the numbers and you have: +1.00 = 1/1

D = Diopters

d = distance from lens to focal point, in meters

This tells us is that a plus lens that reaches a point focus 1m away has a diopter power of **+1.00D**.

Diopters = 1/distance in meters

Focal Length in Meters	Formula	Diopters
4m	D = 1/4m	+0.25
2m	D = 1/2m	+0.50
1.33m	D = 1/1.33m	+0.75
1m	D = 1/1m	+1.00
.8m	D = 1/.8m	+1.25
.667m	D = 1/.667m	+1.50
.57m	D = 1/.57m	+1.75
.5m	D = .5m	+2.00
.33m	D = 1/.33m	+3.00
.25m	D = 1/.25m	+4.00
.2m	D = 1/.2m	+5.00
.1m	D = 1/.1m	+10.00

Focal Length Examples – Plus Lenses in Diopters

Visualize the pattern and relationship between focal length and diopter power.

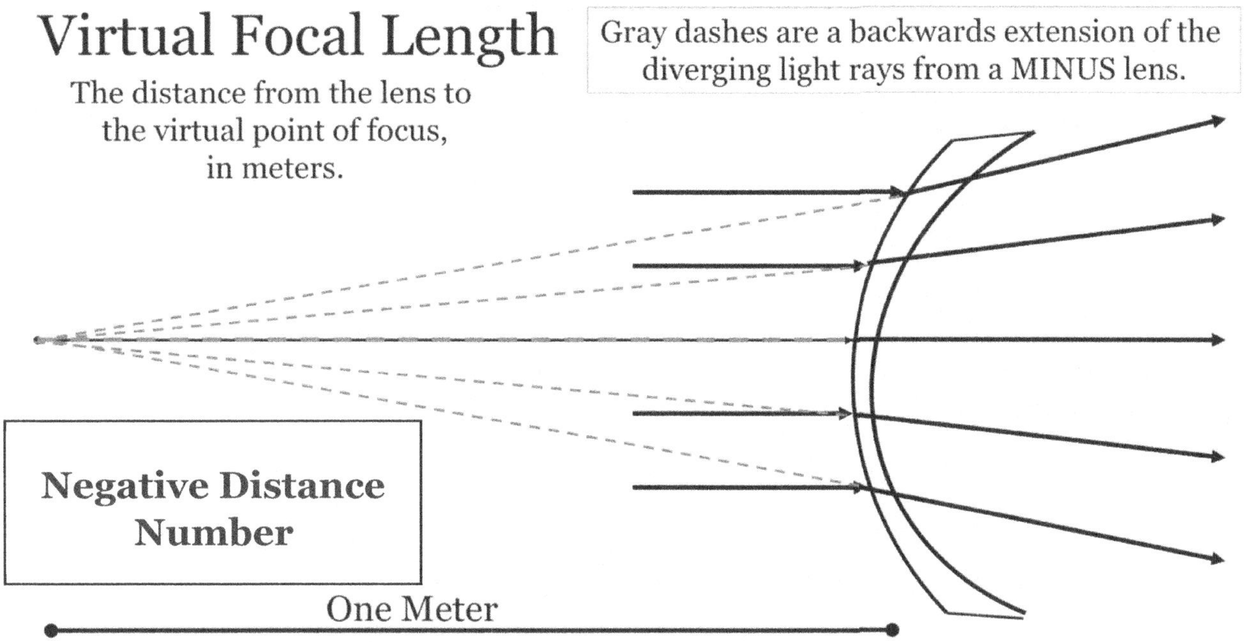

The formula to determine the diopter power is **D = 1/(-d).** Plug in the numbers and you have: -1.00 = 1/-1.

D = Diopters

-d = distance from lens to focal point, in meters

This tells us that a **minus lens** that reaches a **VIRTUAL** point focus **-1m** away has a diopter power of **-1.00D**.

A minus lens does not bring light to a point focus like a plus lens does. The purpose of a diverging or minus lens is to refract light rays apart. To calculate the diopter power of a minus lens we must use a VIRTUAL focal length. To do this we diagram the diverging light rays from the minus lens and then trace backwards to the point that the imaginary focal point would exist. From that the math is identical to the formula used for plus lenses with the addition of a negative distance measurement.

Virtual Focal Length Examples

Minus Lenses in Diopters Diopter Power

Negative Focal Length	Formula	Diopters
-4m	D = 1/-4m	-0.25
-2m	D = 1/-2m	-0.50
-1.33m	D = 1/-1.33m	-0.75
-1m	D = 1/-1m	-1.00
-.8m	D = 1/-.8m	-1.25
-.667m	D = 1/-.667m	-1.50
-.57m	D = 1/-.57m	-1.75
-.5m	D = 1/-.5m	-2.00
-.33m	D = 1/-.33m	-3.00
-.25m	D = 1/-.25m	-4.00
-.2m	D = 1/-.2m	-5.00
-.1m	D = 1/-.1m	-10.00

In college physics you could spend a semester on real and virtual images. An optician just needs to know they exist for certification. We can leave most of the math to the engineers.

The primary difference between image types is whether light rays actually meet or just appear to meet (Walker, 2014).

Real Images

Light rays physically converge, or "cross over," at a specific point in space (Serway & Jewett, 2018). You could capture a real image on a piece of paper or a screen (Hewitt, 2014). These images *almost always* appear upside down (Hewitt, 2014).

Virtual Images

Light rays spread apart (diverge), so they never actually touch (Serway & Jewett, 2018). Your brain traces the rays backward to an imaginary meeting point (Hewitt, 2014). You cannot catch this image on a screen because the light isn't truly there (Walker, 2014). Virtual images are always right-side up (Hewitt, 2014).

Lens Power Formula and the Lens Clock

With the use of a Lens Clock an optician can find what is called the **Nominal Power** of a lens. This measuring gauge can be held against both surfaces and the difference between the convex and concave surfaces gives you an *approximate* power. There are variables such as lens material and thickness that make this a close, but inaccurate method. Still, it is useful as a rule of thumb measurement.

D1 = Front Surface Power in Diopters

D2 = Back Surface Power in Diopters

D = Nominal Lens Power in Diopters

D1 + D2 = D Nominal

For example:

The lens clock reads a front surface of +10

The back-surface reads – 2

+10.00 + (-2.00) = +8.00D

The front surface power of a lenses is also called the **Base Curve.** Typically, the software of the lens manufacturer will select the proper base curve for the power of the lens. However, there will be times when you want a special base curve or to match the base curve the patient has been using (an important part of troubleshooting). The lens clock will give you that front curvature and that can be specified when you order or select a lens.

Vertex Distance

"Vertex distance is the distance between the back surface of a corrective lens and the front of the cornea…Increasing or decreasing the vertex distance changes…the power of the lens relative to the eye…" (Vertex Distance, n.d.).

Vertex distance is the gap between the back of the eyeglass lens and the front of the eye. Light has to travel through that gap after it leaves the lens but before it enters your eye.

If a patient changes the distance of that gap by sliding their glasses down their nose or pushing them closer, it change the effective power of the lenses. Moving the lens further away or closer makes the "power" of the glasses feel stronger or weaker, even though the lens itself hasn't changed. This is why **fitting frames correctly is just as important as the prescription itself**.

KEY VERTEX DISTANCE TAKE-AWAY

Lens Farther Away	More Plus	Less Minus
Lens Closer	Less Plus	More Minus

Take a close look at the basic math as it relates to lens position. Know that vertex distance changes the resulting diopter power. **Handy to know when troubleshooting**.

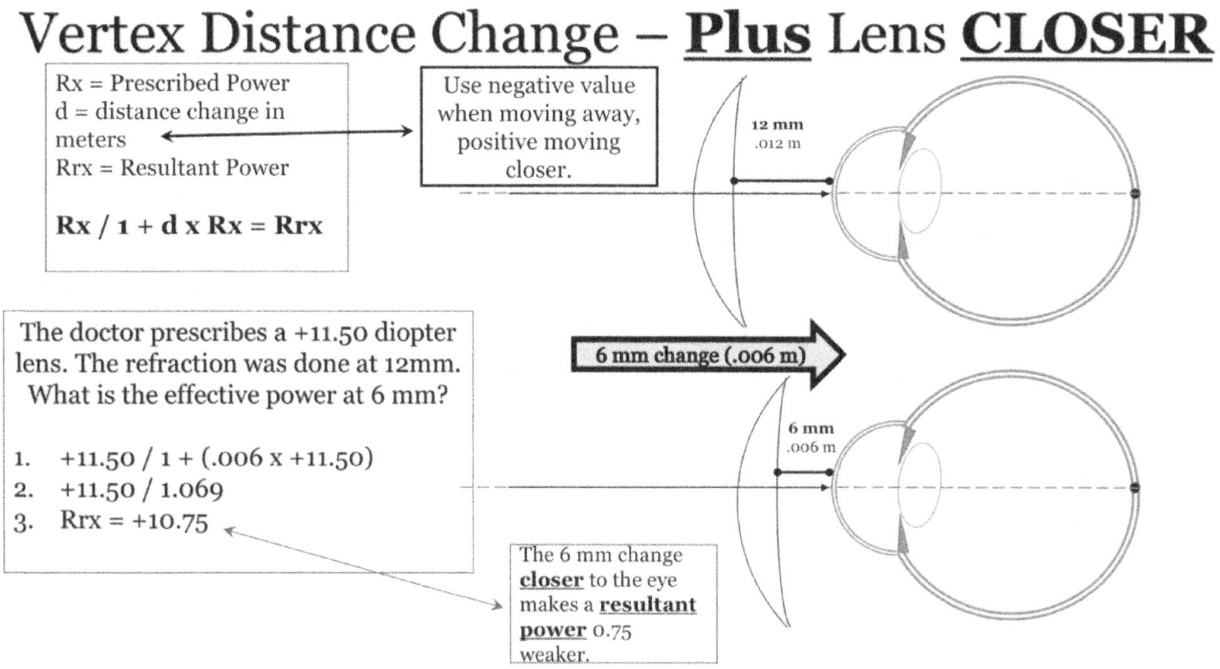

Vertex Distance Change; Plus Lens

➢ *Distance Effects Diopter Power*

Closer – Less Magnification
Plus Lens Weaker

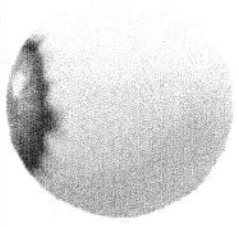

Farther Away – More Magnification
Plus Lens Stronger

Do the math on a separate sheet to lock the concept in your memory.

Vertex Distance Change **Minus** Lens **Closer**

Rx = Prescribed Power
d = distance change in meters
Rrx = Resultant Power

Rx / 1 + d x Rx = Rrx

Use negative value when moving away, positive moving closer.

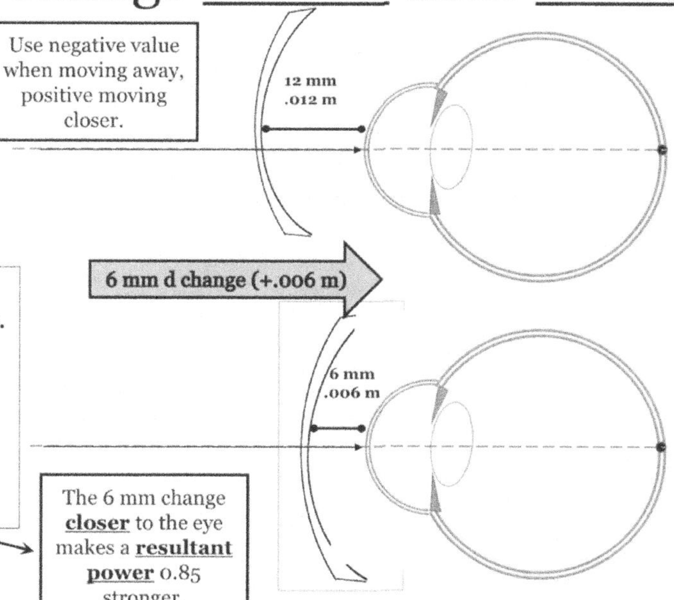

The doctor prescribes a -11.50 diopter lens. The refraction was done at 12mm. What is the effective power at 6 mm?

1. -11.50 / 1 + (.006 x -11.50)
2. -11.50 / 1.069
3. Rrx = -12.35

The 6 mm change **closer** to the eye makes a **resultant power** 0.85 stronger.

Vertex Distance Change **Minus** Lens **Away**

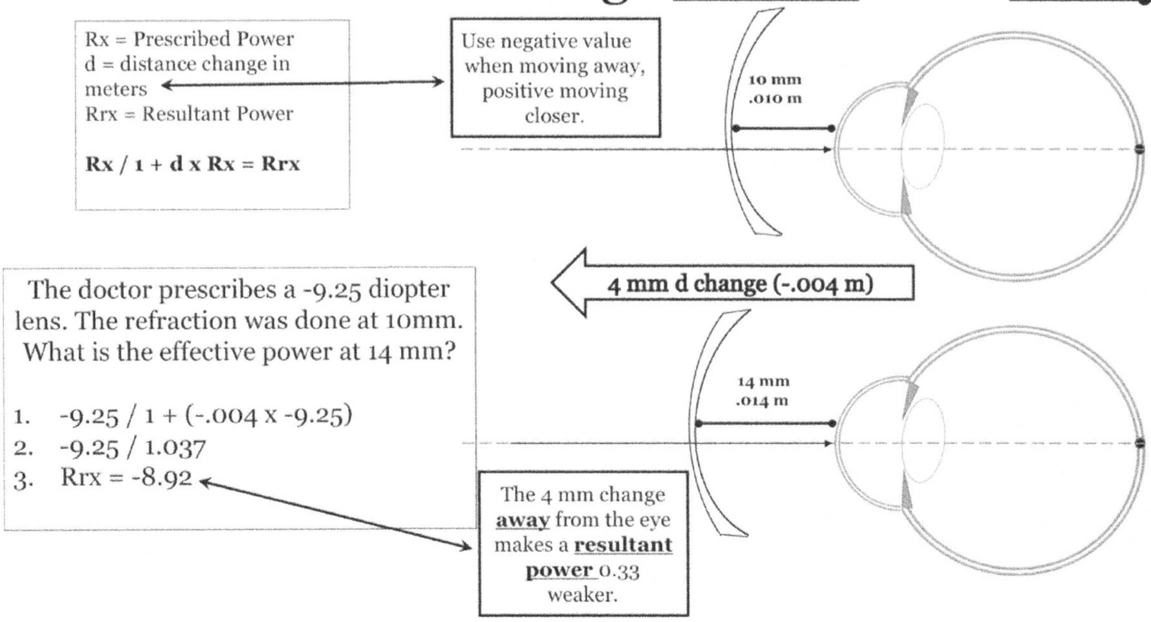

Practical Example: A person wearing reading glasses might slide them down their nose to gain a bit more "boost" for very small print.

Practical Example: A nearsighted person who pushes their glasses tight against their face is often trying to make the correction feel stronger to see distant objects more clearly.

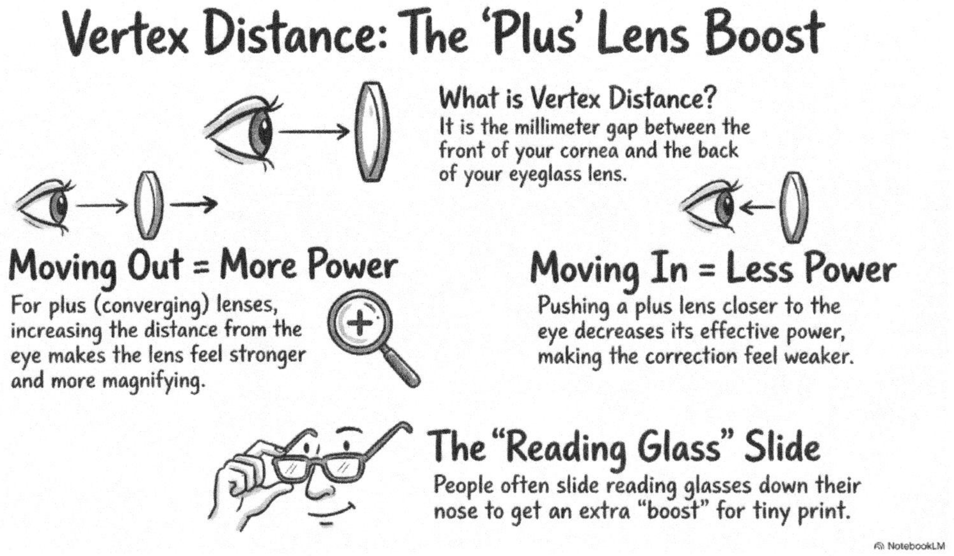

Chapter Seven

Toric lenses and the Optical Cross

EYEGLASS prescriptions start with the spherical component of the prescription. That type of lens is like cutting off a slice of a round ball. But most people have at least a slight astigmatism. Correcting that requires a lens that is not perfectly spherical, like the ball, but must be a toric surface, like a slice off a football

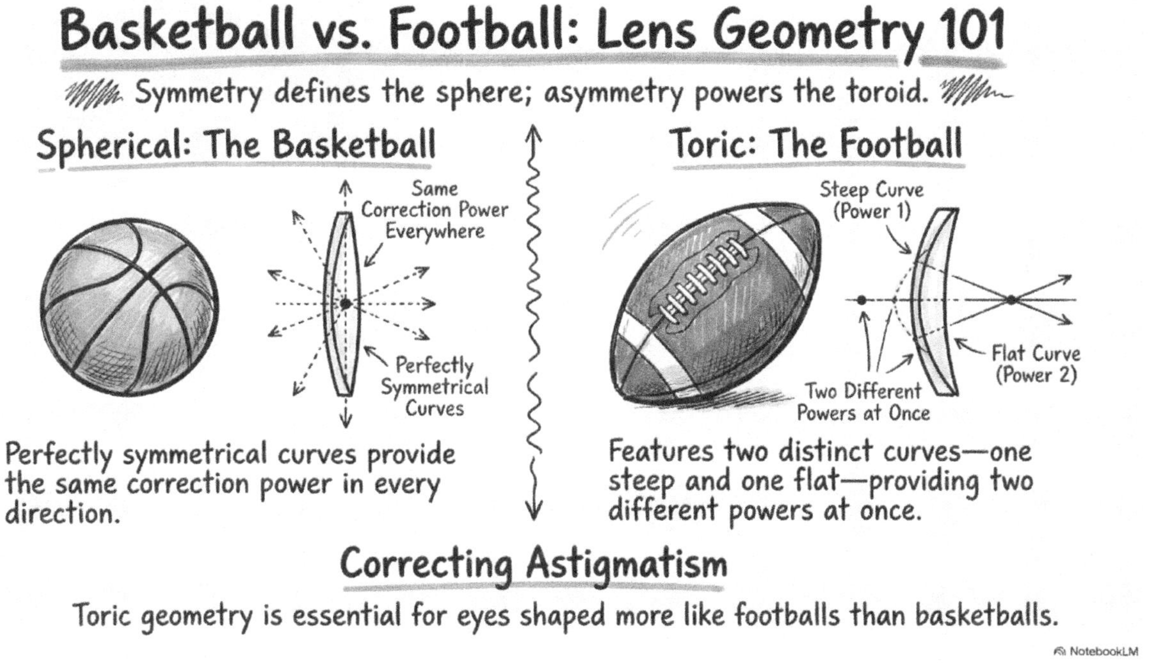

A toric surface is like cutting a slice off the football. This gives you two different prescriptions on the same lens. On one principle meridian you will have the spherical Rx. Perpendicular to that meridian will be the full correction for the astigmatism, a different power on the same lens. A toric lens is one having full sphere power in one meridian and the prescribed astigmatic power 90 degrees away.

As an optician you will learn to recognize the type of prescription often by looking at the lens blank. You know from studying prisms that plus lenses are thicker in the center and minus lenses are thicker on the edges.

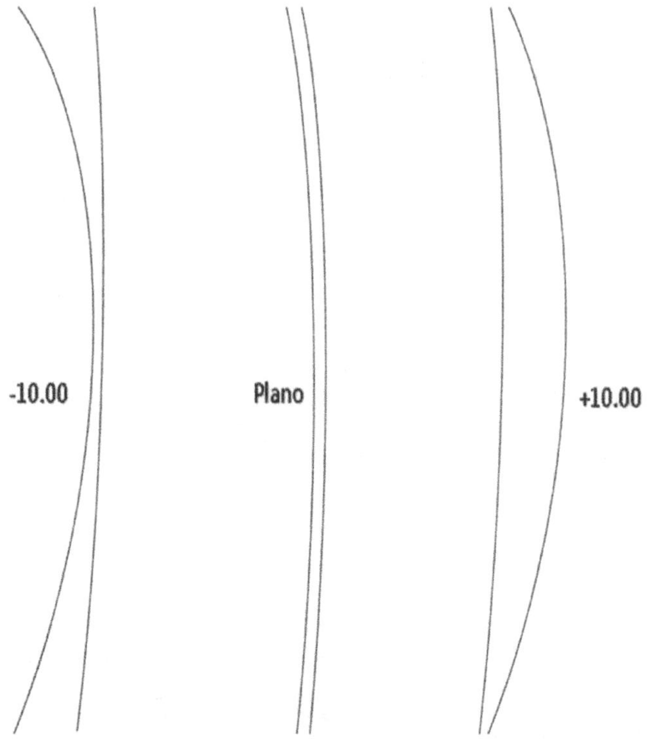

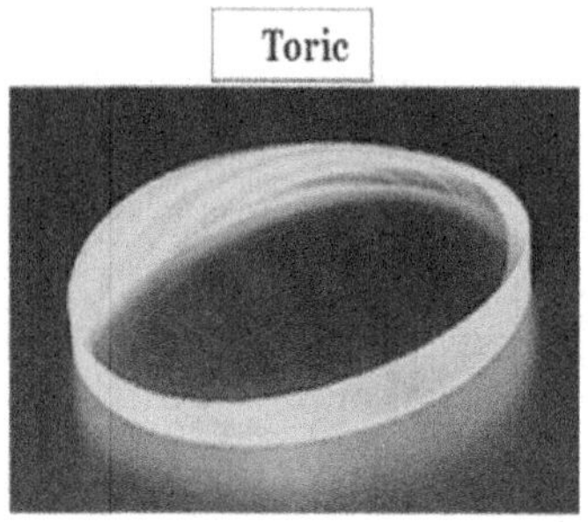

Spherical lenses tend to be generally uniform in center or edge thickness. With a toric lens blank you will observe curvatures more reminiscent of a saddle.

Consider the powers illustrated above, the -10.00 and the +10.00. Imagine, on the cross, that the +10.00 power was on the horizontal line and the -10.00 was on the vertical line. This would be a difference of 20 diopters between the horizontal and vertical meridians. The **vertical** meridian would have the edge thickness of a **-10.00** lens and the **horizontal** meridian would have the center thickness of a **+10.00**. The vertical meridian's center thickness would be dictated by the +10.00. This is a very extreme example, but you can imagine the saddle shaped curve shape the lens blank would take on. Opticians use the optical cross to visualize the doctor's prescription when it includes correction for astigmatism in addition to the spherical power.

Take your time understanding the optical cross. It is a key concept of an optician's skill.

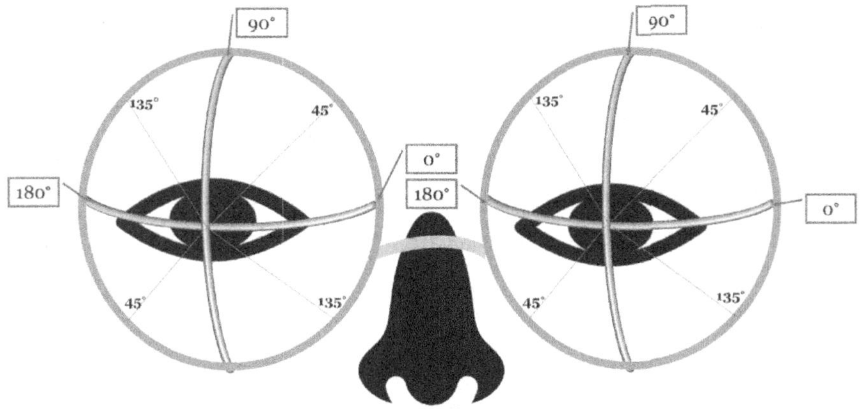

Optical Cross Concept

When you look at the prescription the eye doctor wrote your legal obligation is to align the powers of the lens in front of your patient's eyes in a way that corrects the errors in their eyes. Consider the following Rx:

	Sphere	Cylinder	Axis	Add
OD - Right	+1.00	-1.00	180	
OS - Left	+1.00	-1.00	180	
				PD = 60

Absent the portion of the Rx that requires **cylinder and axis** then both lenses would have a power of **+1.00** in all meridians, vertical (90°), horizontal (180°), and all other meridians. However, in this case, the **cylinder and axis** components are **not optional**. Given that, the only meridian that is +1.00 is the horizontal or zero-degree or 180-degree line (because of the 180-axis prescribed). Note that the horizontal, the zero degree and the 180-degree (on the cross) are all the same thing. They are the same exact meridian.

When you look at this Rx you should understand that it means the following:

1. The power is +1.00 on the horizontal or 180° line.
2.The cylinder power is exactly 90 degrees away from the sphere power.
3.The full cylinder power is the sphere (+1.00) plus the prescribed cylinder (-1.00).
4.Added together the resulting full cylinder power is 0.00 or plano.
5.The power on the vertical or 90-degree meridian is 0.00.
6.The power on all meridians between 180 and 90 gradually and evenly changes. For instance, in this scenario, the power on both the 45° AND the 135° meridian is +0.50 (half the cylinder).

Another way to read each line is:

The power is **+1.00** on the 180 line and **-1.00** is added to the sphere power 90° away and that equals a power of **0.00**.

Once you have the basic concept down you can understand what is needed even if the Rx is written differently. The example above is written in "**minus cylinder form**". Some doctors will write the Rx in "**plus cylinder form**". That form has a plus sign in front of the cylinder. **The process and the math are the same**. Do not let it throw you off. Practice will make it easy. More on this to come.

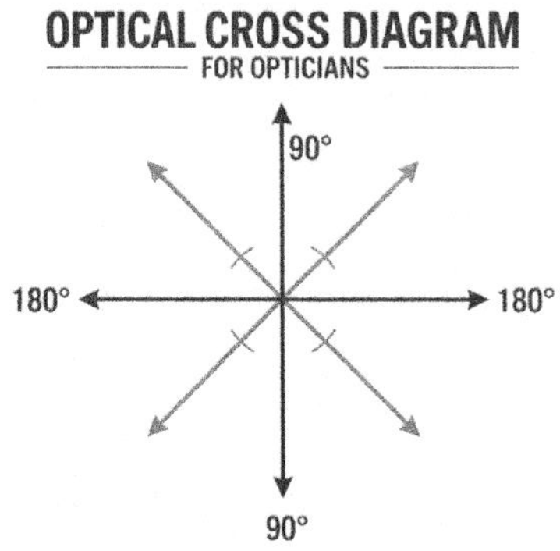

Chapter Eight

Anatomy of the Eye Human Eye

ESSENTIAL ANATOMY OF THE
HUMAN EYE

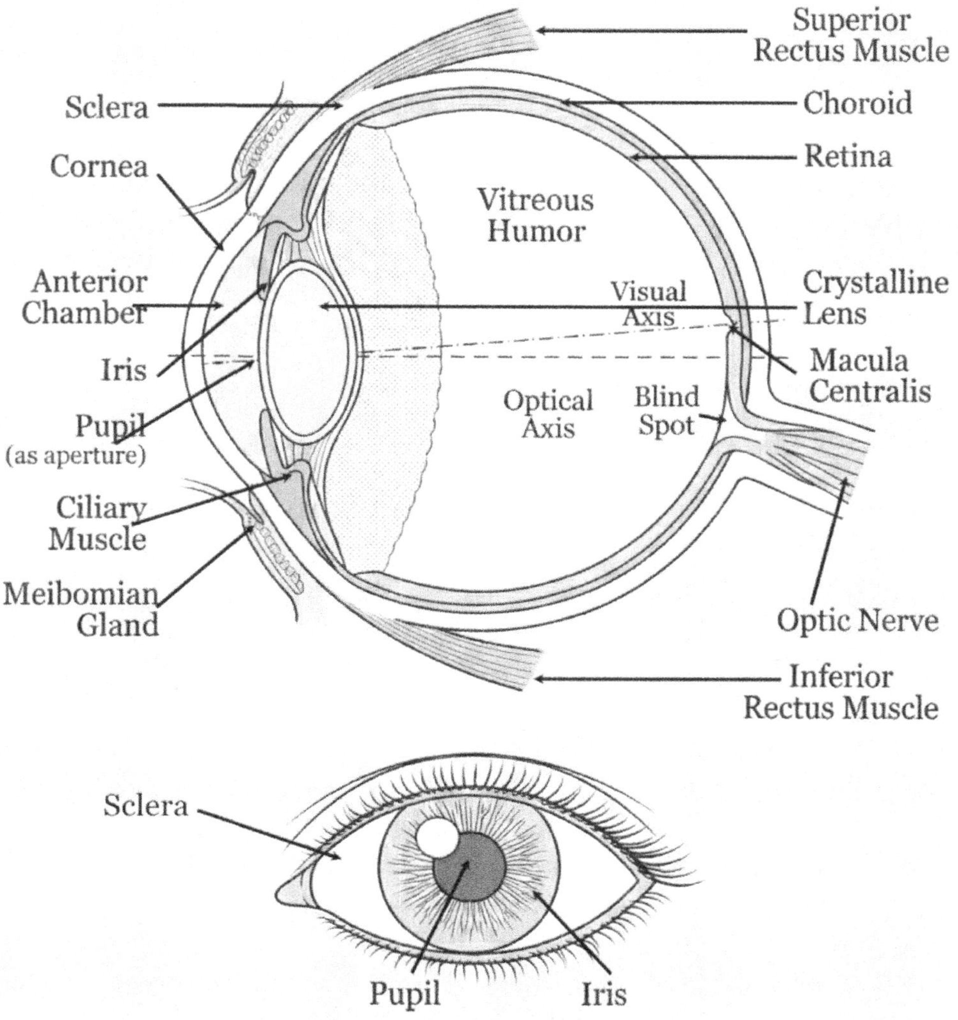

Anatomical Front View

THE human eye functions as a sophisticated optical gateway, transforming light waves into electrical signals through a complex network of refractive and neural components. Each structure plays a specific role in maintaining the clarity and focus of visual perception.

Elements of Eye Anatomy

Cornea: The clear, dome-shaped outer layer at the front of the eye that protects it and helps focus light as it enters. The cornea provides the majority of the refractive power of the eye as it has, on average, about 40 to 45 diopters of power.

Sclera: The tough, white outer part of the eye that provides structure and protects the inner parts.

Iris: The colorful part of the eye that controls how much light gets in by changing the size of the pupil in response to lighting levels. It is larger opening in the dark, smaller when bright. *

Pupil: The black opening in the center of the iris that lets light through to the back of the eye.

Crystalline Lens: A clear, flexible structure behind the pupil that changes shape to help the eye focus on objects at different distances. This flexible lens provides 16 to 20 diopters of plus power with an additional focusing strength of up to 10 more diopters during youthful years. This power reduces with age.

Ciliary Muscle: A tiny muscle that pulls on the lens (or relaxes) to change its shape so you can focus on things up close or far away.

Aqueous Humor: A clear, watery fluid in the front of the eye that helps maintain its shape and provides nutrients.

Vitreous Humor: A clear, jelly-like substance that fills the large space in the back of the eye to help it keep its round shape.

Retina: The thin layer of light-sensitive tissue at the back of the eye that acts like film in a camera to capture images. It transmits data to the optic nerve from the rods and cones.

Macula Fovea: A small, special area in the middle of the retina that allows you to see fine details clearly. The densest area of photoreceptors, mostly cones.

Optic Nerve: The thick cable of nerve fibers that carries visual information from the retina directly to the brain.

Blind Spot: The specific area on the retina where the optic nerve leaves the eye; because there are no light-sensitive cells here, it cannot see light.

Think of the elements of the eye as you label and/or color in the parts here:

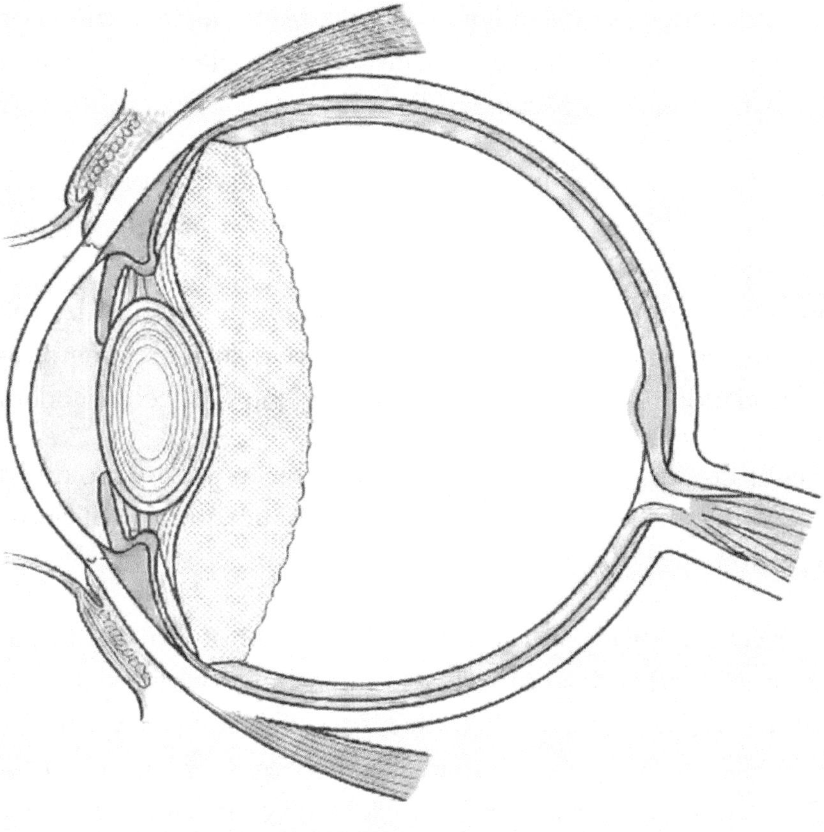

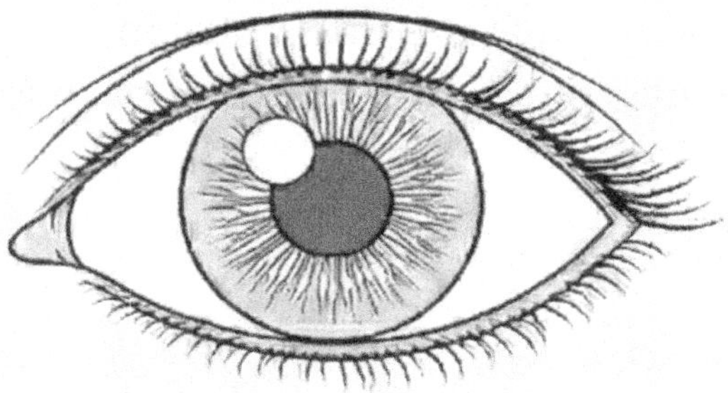

Need-to-Know Anatomy Concepts

The human eye has many similarities to a camera although quite a bit more miraculous. Light enters the first lens, the cornea. The cornea's curvature provides 40 to 45 diopters of plus power. You will probably never make a pair of glasses with 45 diopters of plus power. It is quite strong. It begins to converge light rays on its journey to the back of the human camera.

Our light rays travel through the anterior chamber and the clear fluid called aqueous humor.

Light rays then pass the structure called the iris. This marvelous structure opens and closes the pupil to control the amount of light that enters the inner portions of the eyeball. The iris, controlling the pupil size, also serves to cause the pinhole effect. That effect allows light rays that are parallel, as opposed to scattered light rays, to proceed to the crystalline lens.

The flexible crystal clear biconvex crystalline lens serves more than one purpose. With a diopter power of roughly 16 to 20 it continues to converge light rays to focus on the back of the eye. In addition, this lens, suspended by a ring of ligaments (Ciliary Muscle) can fine tune the focus by becoming less biconvex or more biconvex as needed. When we look at objects that are close the lens thickens to a greater biconvex form creating more plus power. This is called accommodation. Whenever you are reading you are accommodating.

By now the light rays that have reflected off the objects we are looking at are converging to a point focus with the intent to land on the greatest concentration of photoreceptor cells on the back of the eye, the macula fovea.

Photoreceptor cells, the rods and cones, are like the photosensors in your cell phone camera. Cones detect color and rods are the black and white receptors. They are spread across the inner layer of the back of the eye called the retina. The cones have their greatest concentration at the fovea while the rods predominate on the edges of the retina. The retina receives its blood supply from the choroid layer. The sclera, or white of the eye, is the tough outer layer.

All the data collected by photoreceptors is transmitted to the brain via the optic nerve. This nerve transmits the data to the visual cortex, an area of the brain that interprets the data and forms the images we call vision.

The Blind Spot

The visual cortex, in the back region of the brain, can do some amazing things to make the images we see seamless and vibrant. Where the optic nerve leaves the retina (back of the eye) is call the blind spot. This is because no photoreceptors exist on this spot. So why don't we see a missing spot in our vision? Because the visual cortex "paints" in the missing data to form a smooth image. Try this experiment to gain a greater understanding.

You will need an 8 by 11.5 piece of paper and a metric ruler.

On your paper draw a circle 1 centimeter (cm) in circumference and fill it in, Then, 13 cm to the right draw an X like the illustration

Hold the paper about 45 cm away, stare at the X and close your right eye. Move the paper closer and then farther away. When the image lands on your optic nerve, or blind spot, the dot will disappear. That image of the dot disappears because there are no photoreceptors on the retina where the image lands. Open your other eye and the dot reappears because photoreceptors from the other eye help "fill in" the blind spot.

The Optic Nerve

The optic nerve is one of the twelve pairs of cranial nerves. "Cranial nerves relay information between the brain and parts of the body, primarily to and from regions of the head and neck, including the special senses of vision, taste, smell, and hearing."

The **optic nerve** acts as a high-speed data cable that links your eye directly to your brain, serving as the most vital sensory bridge in your body (Hewitt, 2014). This sophisticated structure functions as a direct extension of your central nervous system rather than a simple

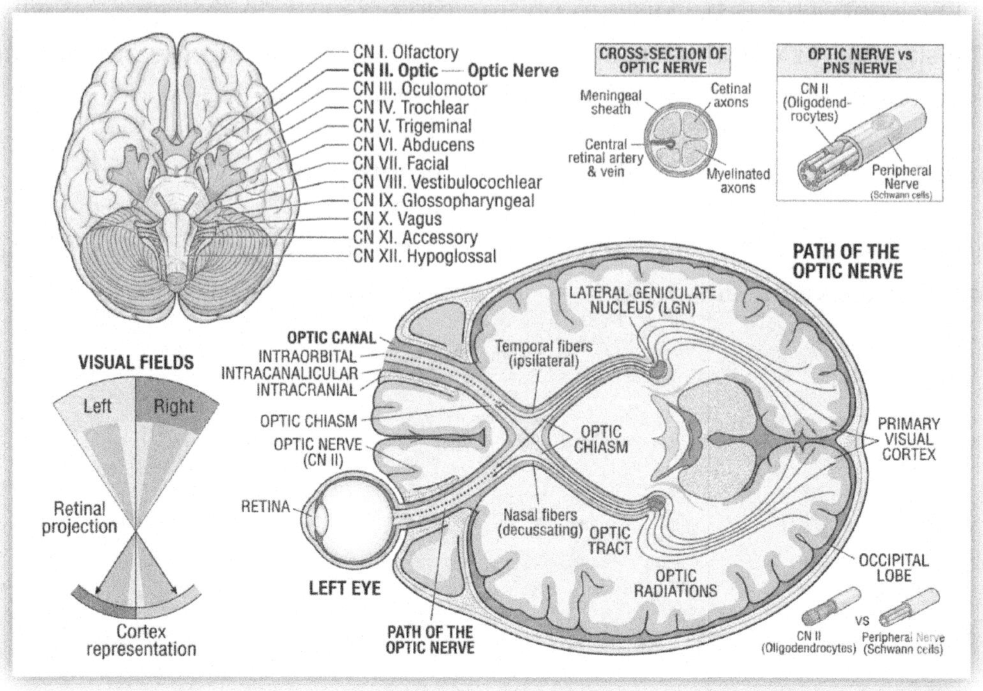

peripheral wire. It transmits the energy of light, the photons that strike the retina and are converted into electrical signals, to the visual processing portion of the brain.

The **Optic Nerve** is **Cranial Nerve II**; it belongs to an elite team of twelve nerves that manage everything from your sense of smell to your facial expressions. While the optic nerve handles the "input" of sight, its teammates—the oculomotor, trochlear, and abducens nerves—control the "output" by moving your eye muscles so you can track a flying bird. They work together. The coordination creates visual clarity. Without this constant mechanical teamwork, your vision would be completely static and useless.

The pathway follows a strict four-stage journey that starts at the back of your retina and terminates in the brain's **visual cortex** (Walker, 2014). First, the nerve travels through the eye socket, then through a tiny bone tunnel in your skull, and finally into the brain's main headquarters. It crosses over at the **Optic Chiasm**. Data swaps sides. This ingenious geometric design ensures that both sides of your brain receive a complete, balanced picture of the world around you.

The Muscles of the Eye

If you have ever "rolled your eyes" at a comment, then all the muscles of your eye have played a part. Prior to naming these muscles the **directional terms** used in medicine warrant some review as they are a part of the muscle names.

Superior – ABOVE

Inferior – BELOW

Anterior – BEFORE (to the front)

Posterior - AFTER (to the back)

Lateral – OUTER (to the side)

Medial – INNER (towards the middle)

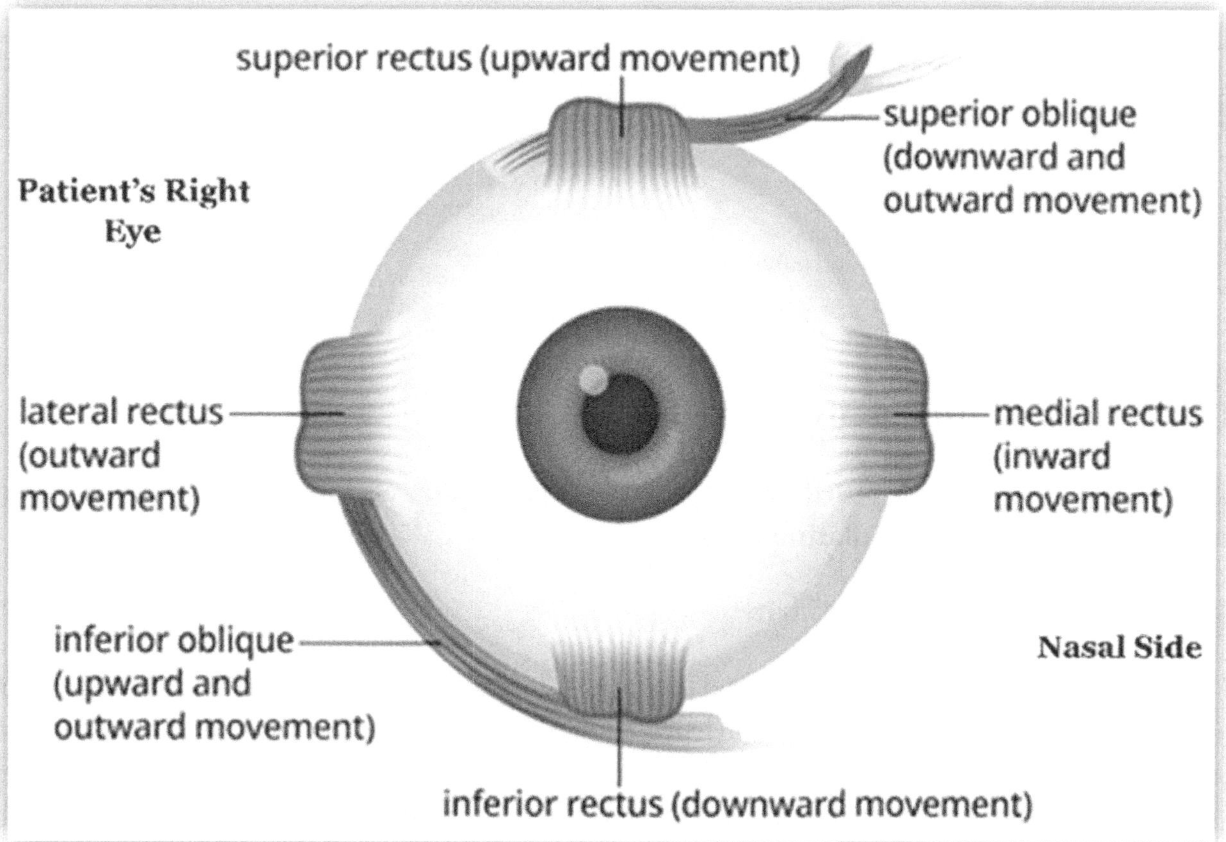

Familiarity with the names and functions of the six eye muscles will aid in your quest for certification.

Muscles, cont.

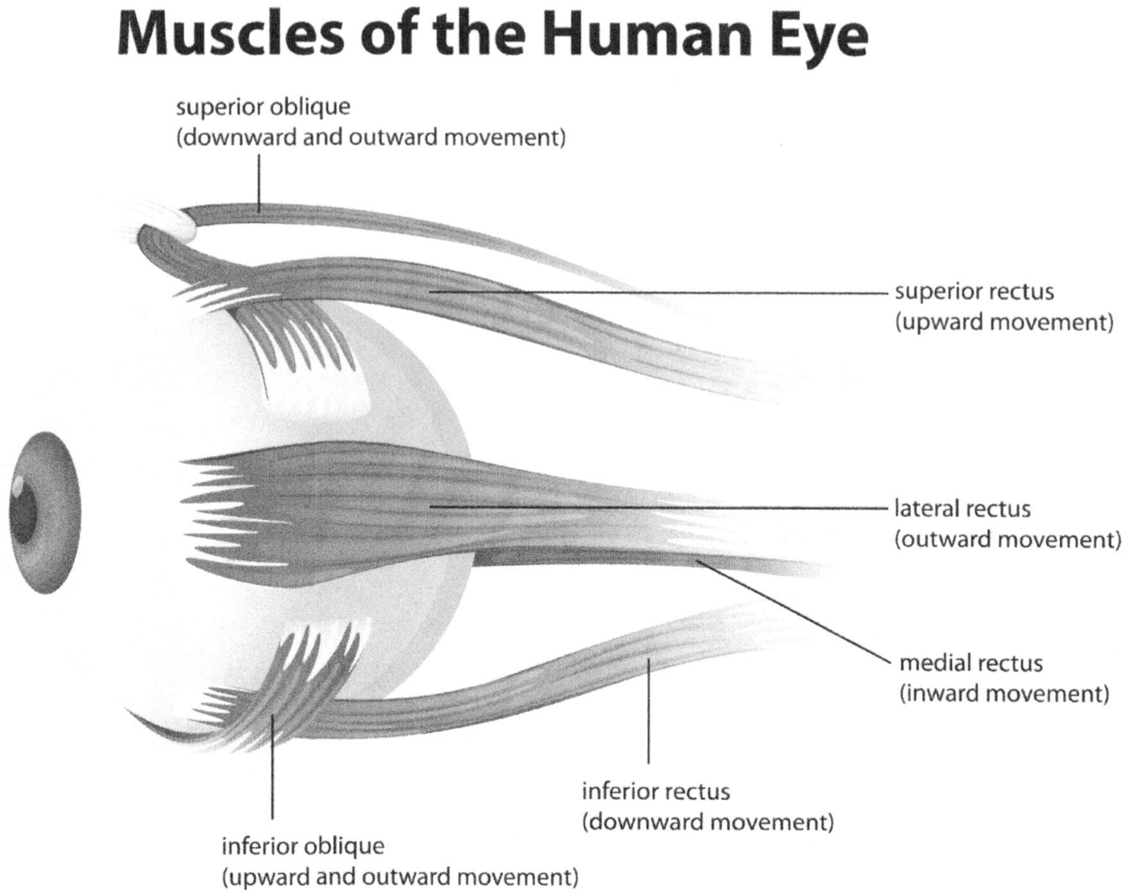

The Bones of the Orbit of the Eye

The orbit of the eye is composed of seven bones: the **frontal, zygomatic, maxilla, ethmoid, lacrimal, sphenoid, and palatine bones**.

Introduction to the Orbital Cavity

While we often think of the area around our eyes as a simple opening, it is technically known as the **Orbital Cavity** (or "orbit"). Far from being a hollow void, the orbit is a sophisticated, bony house meticulously constructed from seven distinct pieces. This architectural marvel must balance two extremes: it features heavy-duty structural pillars designed to withstand impact, alongside paper-thin "partition walls" that separate the eye from the surrounding sinuses.

The Seven Bones

The orbit is a masterwork of biological engineering, utilizing seven different bones to form its walls, floor, and roof. The following table identifies these components and their primary roles in the structural integrity of the "blueprint."

Bone Name	**Primary Location/Wall**	**Key Structural Contribution**
Frontal	Superior Wall (Roof)	Forms the sturdy anterior (front) ceiling of the orbital house.
Zygomatic	Lateral Wall & Floor	Acts as the primary "pillar" for outer structural support and protection.
Maxilla	Inferior Wall (Floor)	Provides the foundational base of the orbit's anterior structure.
Ethmoid	Medial Wall	Serves as a delicate internal divider between the eye and the nasal cavity.
Lacrimal	Medial Wall	A small, specialized component integrated into the inner wall.
Sphenoid	Posterior Part	Forms the deep "back wall" and anchors the orbit to the skull.
Palatine	Floor	Contributes a minor, specialized segment to the posterior floor.

Key Insight: The human eye socket is not a single bone, but a complex "orbital cavity" formed by the convergence of seven unique bones. By working together, these structures transform a simple opening into a highly specialized house that protects, supports, and provides vital lifelines for the sense of sight.

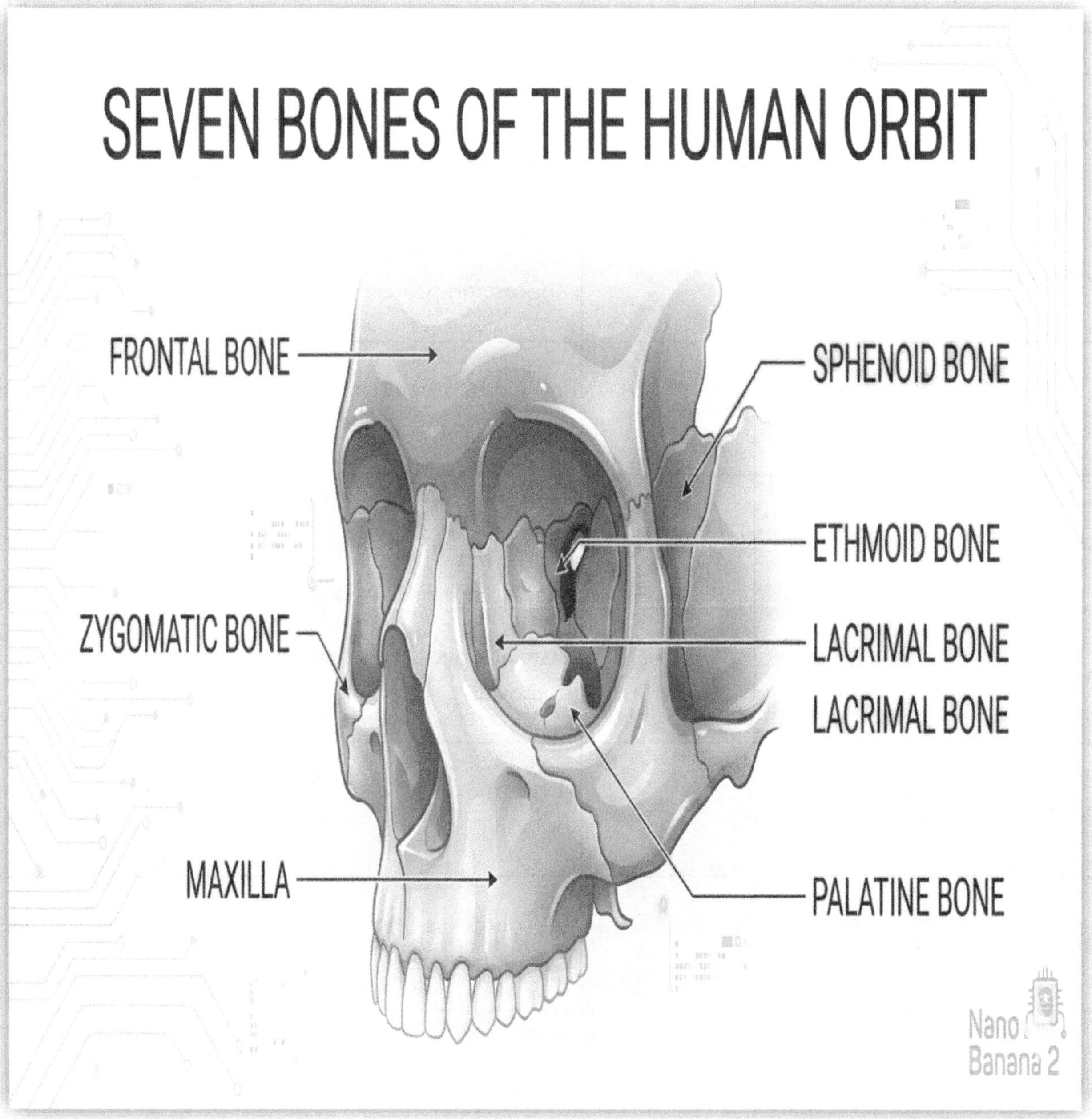

Think of the bones of the orbit of the eye as you label and color here.

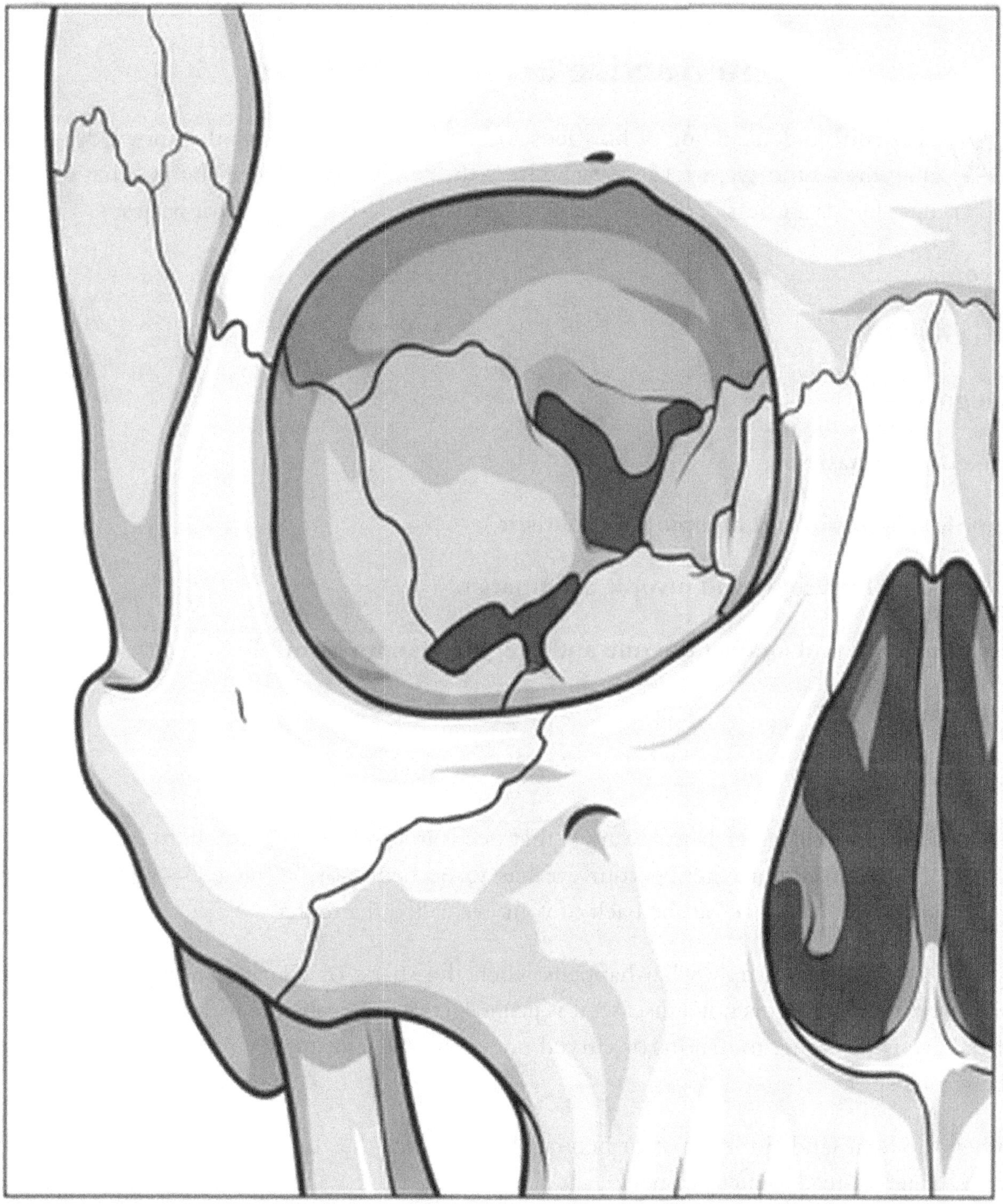

Chapter Nine

Refractive Errors of the Eye

WITH your understanding of how lenses bend light, coupled with the knowledge of the anatomy of the eye, we can now address the refractive errors of the eye that eyeglass and contact lenses address. You will encounter, daily, these refractive error issues:

Myopia

Hyperopia

Astigmatism

Mixed astigmatism

Simple hyperopic and myopic astigmatism

Compound hyperopic and myopic astigmatism

Oblique, with and against the rule and irregular astigmatism

Presbyopia

Cataract (post)

Think of your eye like a high-tech camera that needs to capture perfectly sharp pictures of the world. To do this, light entering your eye has to be bent—or "refracted"—so that it lands exactly on a tiny "bullseye" at the back of your eye called the retina.

A refractive error is simply what happens when the shape of your eye prevents light from hitting that bullseye. It is not a disease; it is just a "geometry problem" with the eye's anatomy. If the eye is too long, too short, or curved unevenly, the light misses the target and lands in the wrong spot.

When the light lands in front of or behind the bullseye, your brain receives a fuzzy, out-of-focus image instead of a clear one.

Glasses and contact lenses are like "corrective shields" that bend the light before it even touches your eye. They adjust the path of the light so that, despite your eye's unique shape, the critical light rays lands righ on the bullseye

Myopia

Myopia, which also called **NEARSIGHTED**. is an error that typically allows the patient to see up close, but distance vision is compromised. Those that are nearsighted **CAN SEE NEAR**.

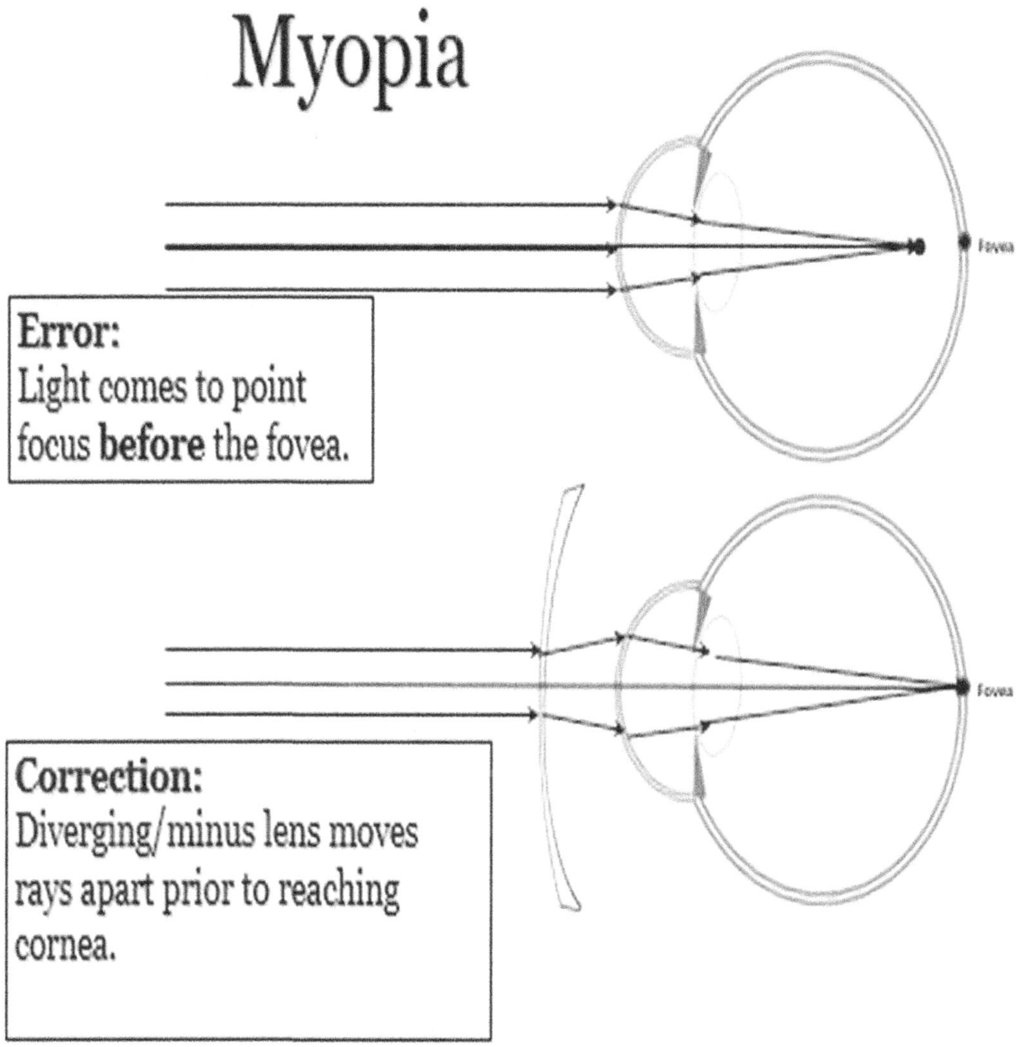

Hyperopia

Hyperopia, also called **FARSIGHTED** is an error that allows the patient to see far away but not up close. Those that are farsighted **CAN SEE FAR**.

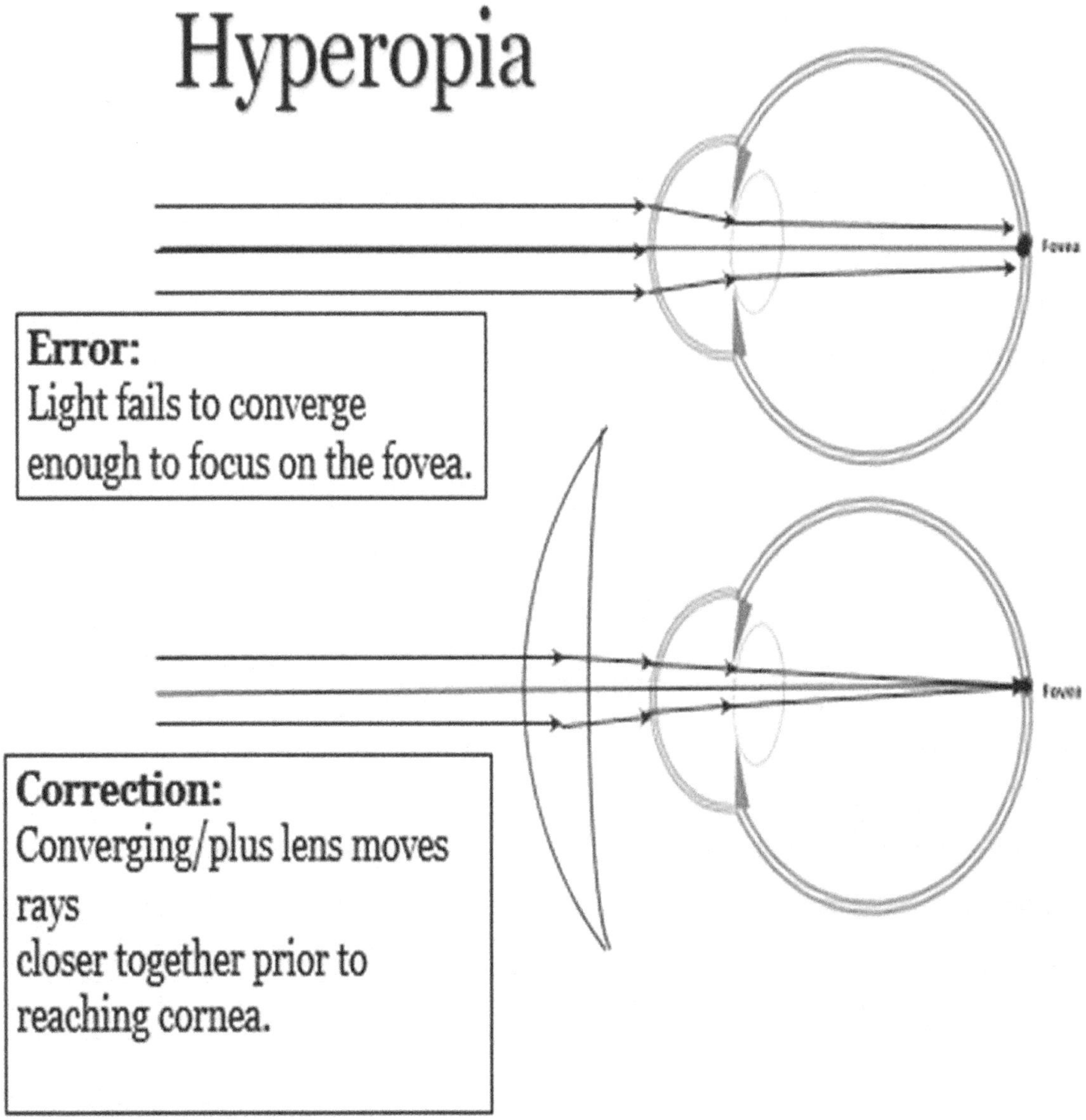

Astigmatism

To understand astigmatism, imagine the cornea—the clear front window of the eye—as a segment of a perfectly spherical basketball. In an ideal eye, this surface possesses uniform curvature in every direction, allowing incoming light to refract evenly toward a single focal point on the retina (Hewitt, 2014).

However, an eye with astigmatism is structurally asymmetric, resembling the elongated curve of an American football. This "toric" shape means the cornea has two different radii of curvature: one relatively steep and one relatively flat (Walker, 2014). Consequently, light entering through the steeper curve bends more sharply than light entering through the flatter one.

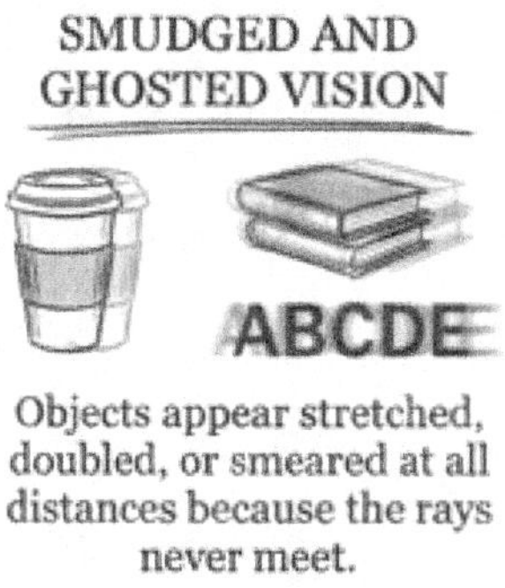

Instead of a crisp point of light, the retina receives two separate focal lines that are perpendicular to each other. This prevents the brain from synthesizing a sharp image, resulting in vision that appears blurred, stretched, or "ghosted" at all distances (Serway & Jewett, 2018).

Corrective lenses for astigmatism are engineered with a specific "cylindrical" power that offsets the eye's unevenness. These lenses are oriented at a precise angle to neutralize the steeper curve, forcing all light rays to finally converge into a singular, coherent point (Hewitt, 2014).

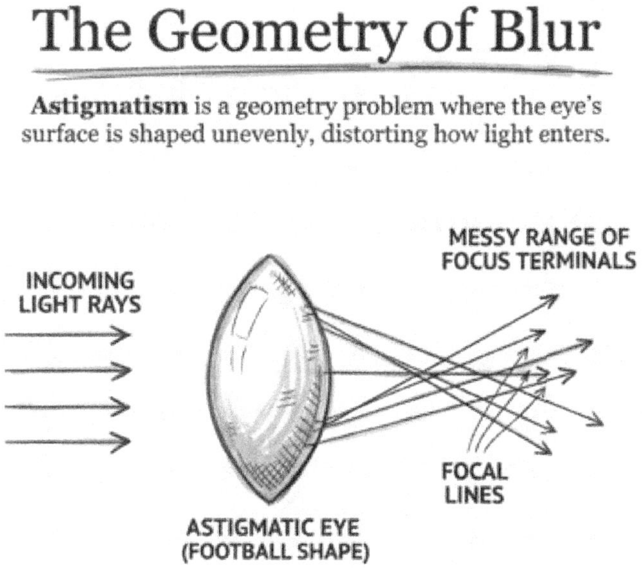

Forms of Astigmatism

Simple Myopic Astigmatism

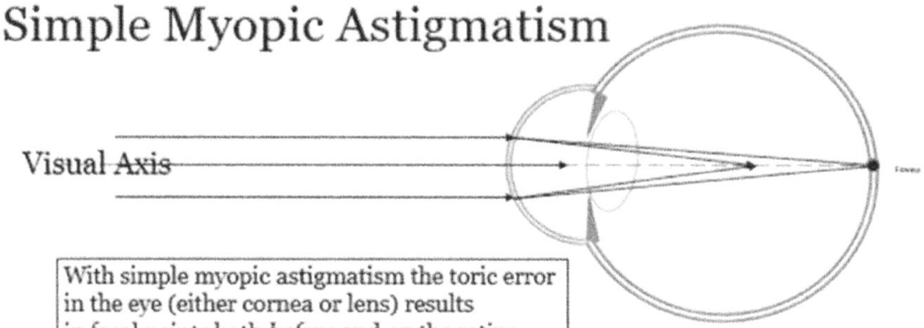

With simple myopic astigmatism the toric error in the eye (either cornea or lens) results in focal points both *before* and *on* the retina.

Compound Myopic Astigmatism

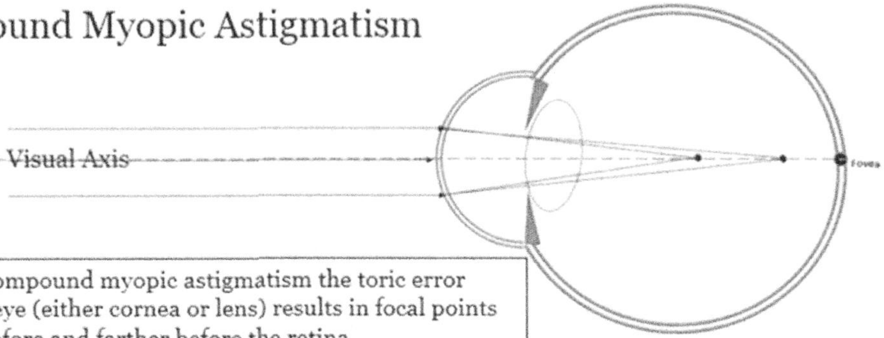

With compound myopic astigmatism the toric error In the eye (either cornea or lens) results in focal points both before and farther before the retina.

Simple Hyperopic Astigmatism

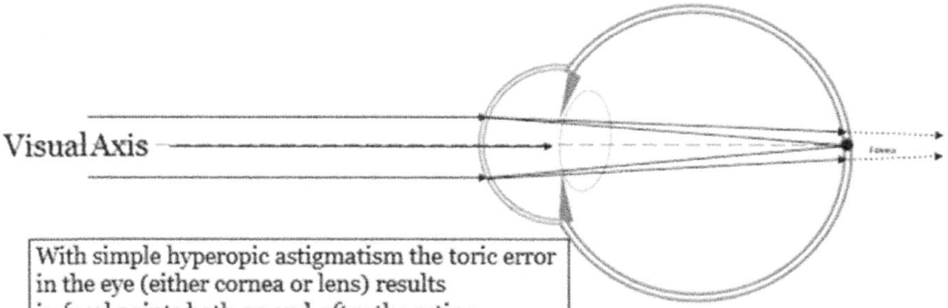

With simple hyperopic astigmatism the toric error in the eye (either cornea or lens) results in focal points both *on* and *after* the retina.

Compound Hyperopic Astigmatism

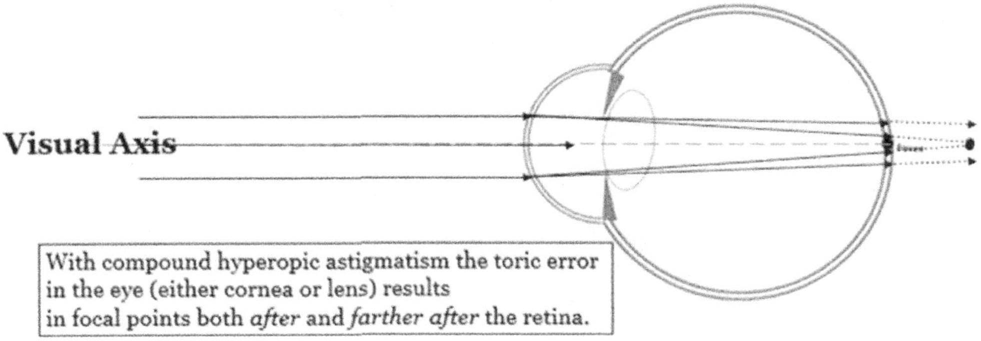

With compound hyperopic astigmatism the toric error in the eye (either cornea or lens) results in focal points both *after* and *farther after* the retina.

Forms of Astigmatism, cont.

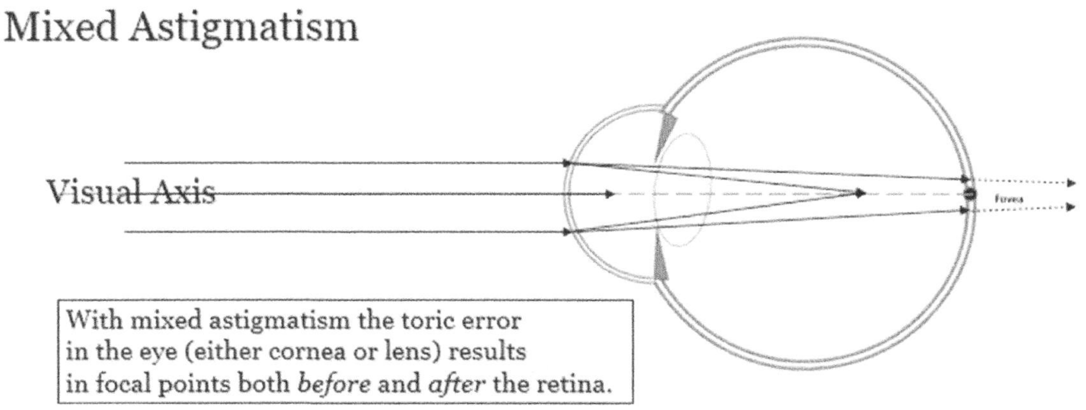

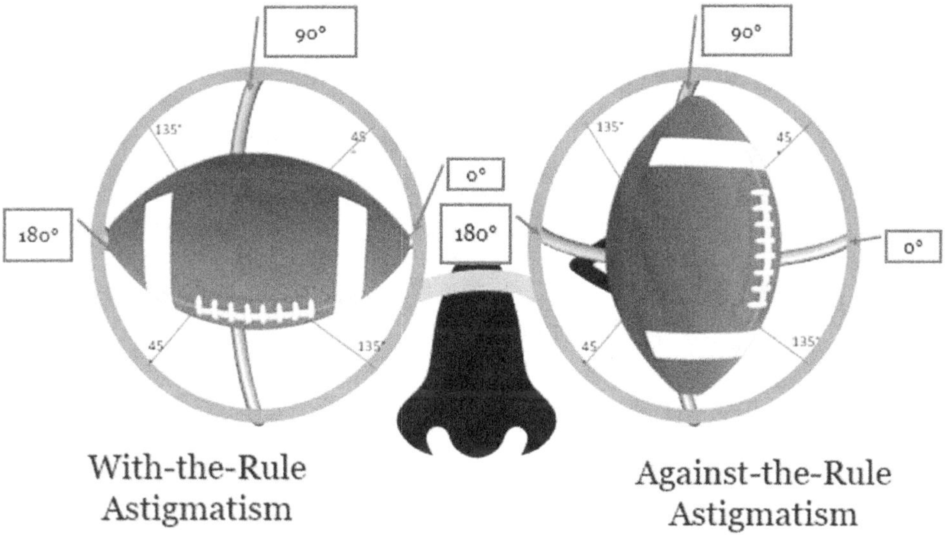

You will hear the terms **with or against the rule astigmatism**. This describes the location of the refractive error on the optical cross.

With-the-rule requires the **flatter meridian is at or near horizontal** (30 to 150 degrees); plus or minus 30 degrees from the 180. **Against- the-rule is vertically oriented** (60 to 120).

Forms of Astigmatism, cont

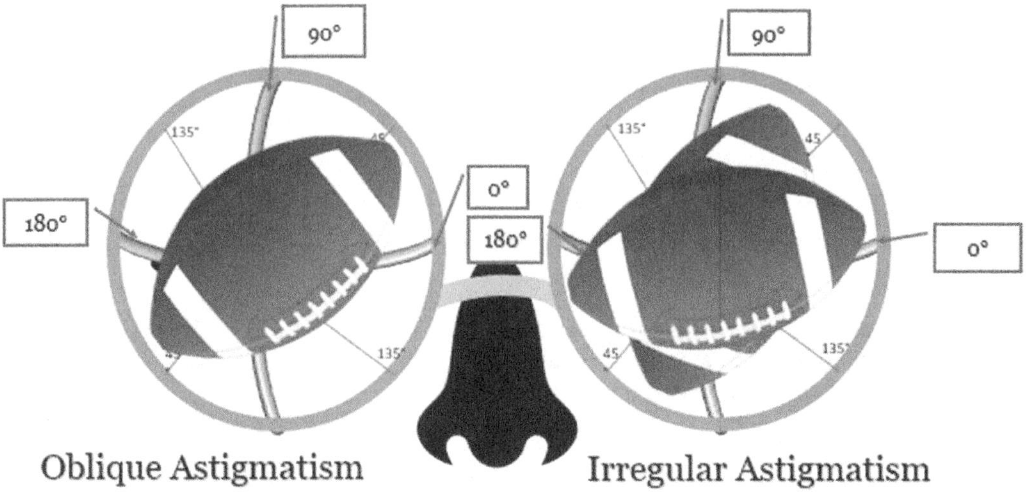

Oblique astigmatism is located **between the with or against** variants. Irregular astigmatism is rare but the toric refractive error is **not separated by 90 degrees** as is normal.

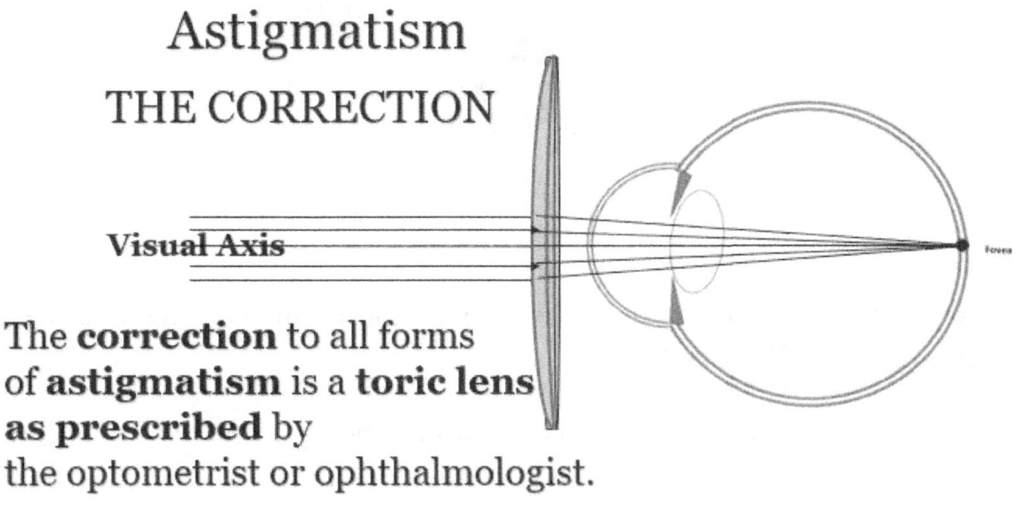

Presbyopia

Presbyopia is a very common physiological shift that occurs as the human eye loses its natural "autofocus" capability with age. To understand this, imagine the crystalline lens inside your eye as a flexible, transparent sac filled with a dynamic protein gel (Walker, 2014). In a younger eye, this lens is highly elastic. Light enters the pupil. Small muscles called ciliary muscles contract to "squeeze" the lens, making it thicker and more curved to focus on close-up objects like a smartphone screen or a book.

This process, known as accommodation, allows for instantaneous focal shifts between the horizon and the hand. Elasticity enables the zoom. As we reach our 40s and 50s, the proteins within the lens begin to denature and cross-link, causing the lens to progressively stiffen (Hewitt, 2014). It essentially becomes a rigid piece of plastic rather than a supple balloon. Even when the internal muscles pull with all their strength, the hardened lens resists changing shape. The "zoom" mechanism fails. Consequently, incoming light rays from near objects cannot be bent sharply enough, and they land in a blurry mess behind the retina instead of directly on it.

This explains why adults often find themselves holding menus at arm's length to find a "sweet spot" where the light still aligns. Geometry dictates the struggle. Correction typically involves external "plus" lenses—found in reading glasses or bifocals—which do the heavy lifting that the internal lens can no longer manage (Serway & Jewett, 2018). These external tools converge the light before it even hits the eye, forcing the image back onto the retinal "bullseye."

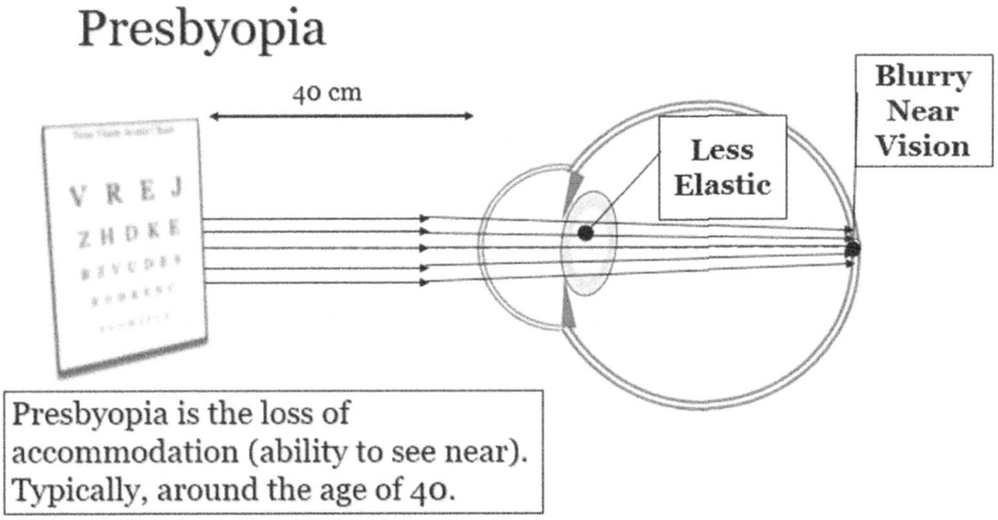

Correcting the diminished flexibility of the internal ocular structure typically involves introducing lenses that converge incoming light rays before they reach the corneal surface (Hewitt, 2014). These reading glasses act as a mechanical substitute for the eye's lost "zoom" capability (Walker, 2014). This adjustment ensures that images of small text land precisely on the sensitive retinal target.

Many individuals opt for multifocal configurations which partition a single lens into distinct zones of varying diopter power to manage both horizon-viewing and near-point tasks simultaneously (Serway & Jewett, 2018). Bifocals possess a visible boundary (Bifocal line) separating two different focal prescriptions (Hewitt, 2014). Progressive Addition Lenses (PAL) hide lines. These seamless designs utilize a vertical corridor of increasing power to mimic natural transitions.

Advanced alternatives include contact lenses that employ monovision, where one eye is calibrated for distance and the other for close-up tasks, requiring the brain to synthesize a clear composite (Walker, 2014). Surgical interventions like replacing the natural stiffened lens with a permanent plastic implant offer long-term clarity (Serway & Jewett, 2018).

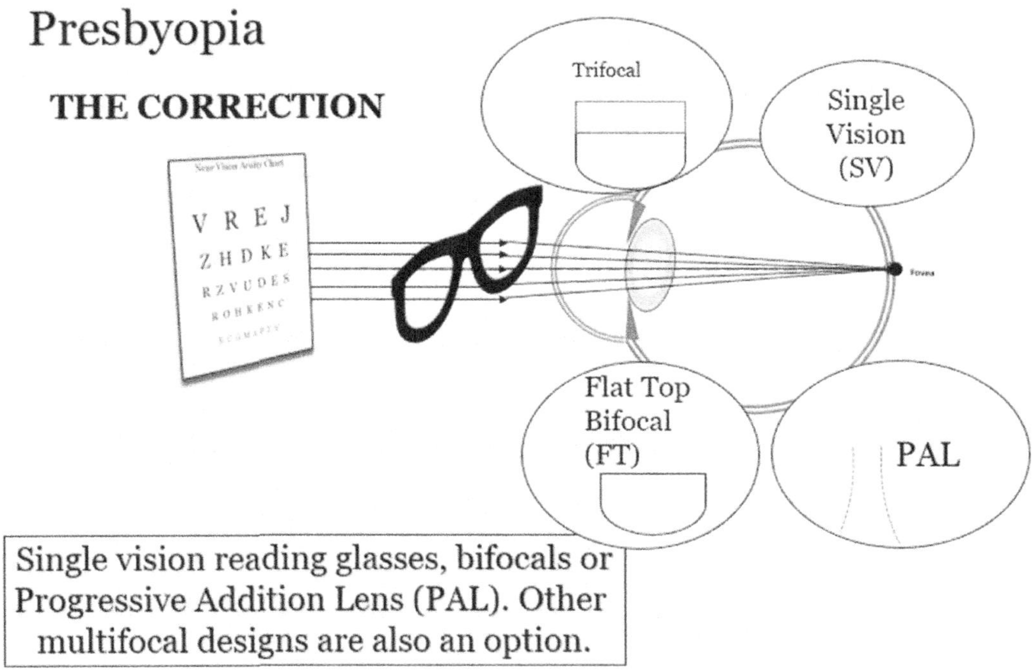

Cataracts

A cataract is the progressive clouding of the eye's natural, internal lens, typically caused by the clumped protein fibers that accumulate over decades of UV exposure or metabolic shifts (Hewitt, 2014). Imagine a clear window slowly turning into a sheet of frosted glass. Light scatters. Instead of a sharp beam hitting the retina, the image becomes dim and yellowed. Other causes are injury or congenital defect.

While lifestyle changes like wearing polarized sunglasses can slow the degradation, surgery is the only definitive correction (Serway & Jewett, 2018). During this procedure, the clouded lens is fragmented and removed. A permanent, medical-grade plastic implant, called an **intraocular lens**, is inserted in its place (Walker, 2014).

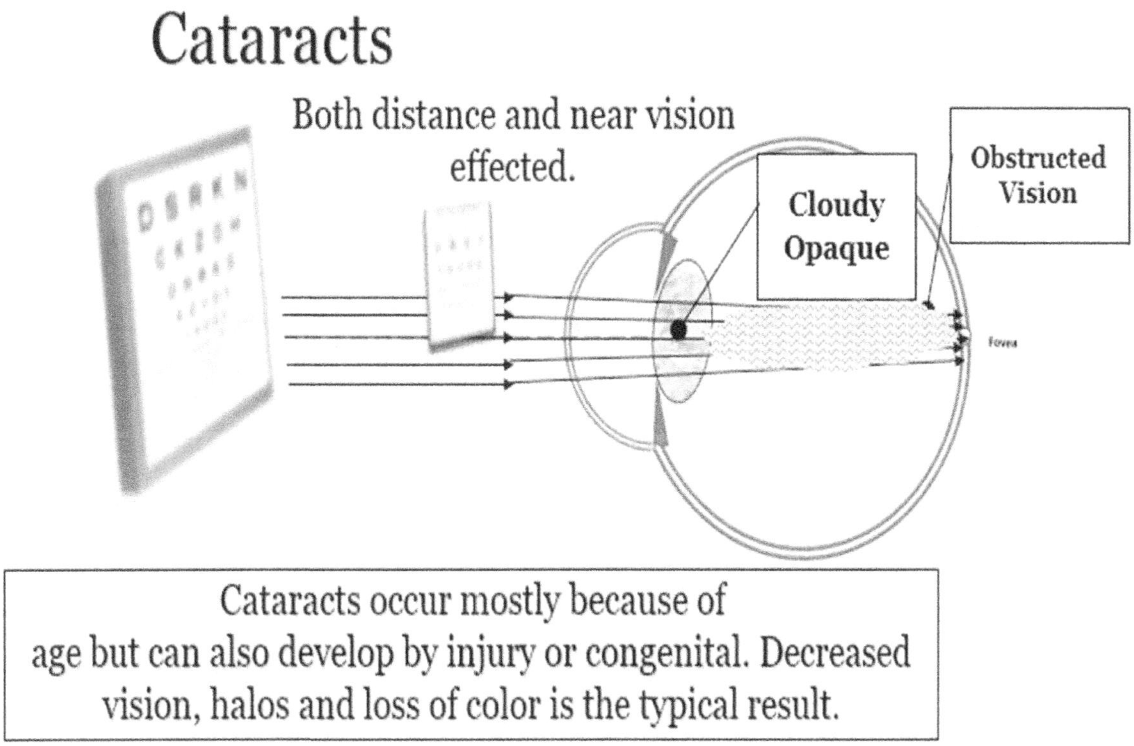

Options for these implants vary based on visual needs:

Single-Focus: Provides crisp distance vision but usually requires reading glasses for close-up tasks.

Multi-Focus: Uses specialized rings to focus light from multiple distances simultaneously, potentially eliminating the need for any glasses (Hewitt, 2014).

Astigmatism-Correcting: Shaped specifically to fix toric eye curves while restoring clarity.

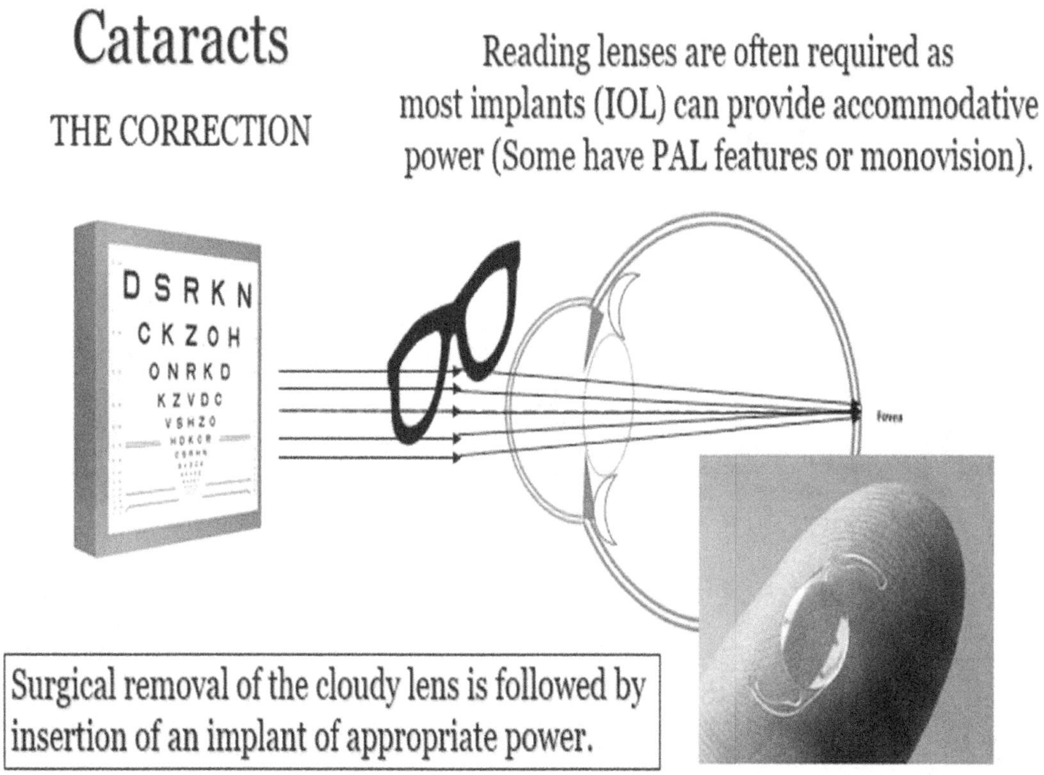

Chapter Ten

Lens Materials

General Characteristics

AN optician has a variety of materials in his/her toolbox all with specific properties that bend light differently. Modern eyewear technology allows individuals to choose lens materials based on a specific balance between visual clarity, physical weight, and impact resistance (Walker, 2014). Each material offers a unique geometric solution to personal vision needs (Serway & Jewett, 2018).

Glass was the original standard for centuries because it provides the most precise "Abbe value" (to be discussed), which means it disperses or scatters light less than any other material (Hewitt, 2014). However, glass is notoriously heavy and dangerous because it can shatter into sharp fragments during an accident. It rarely serves modern users. Safety concerns outweighed clarity.

CR-39 (Plastic) revolutionized the industry by offering half the weight of glass while maintaining excellent optical quality. It remains a popular "budget" choice for many users today. Plastic is softer than glass. It requires a scratch-resistant layer to be applied to survive daily wear and tear but will still scratch without proper care.

Polycarbonate and Trivex are engineered specifically for high-risk environments where eye protection is the primary requirement (Walker, 2014). Polycarbonate is incredibly thin, light weight and impact-resistant, making it perfect for children or sports goggles. Trivex is a premium alternative. It offers the same toughness as polycarbonate but with better optics. Scratch resistance is still a concern and proper care instructions must be part of the dispensing process.

High-Index materials are the ultimate solution for people with very strong prescriptions who want to avoid thick, heavy lenses. These specialized plastics bend light more effectively, allowing the lens to be much thinner and lighter than standard options (Serway & Jewett, 2018). High-index lenses provide the most comfortable experience for complex corrections. But like Trivex, typically comes with premium pricing..

N-Value or Index of Refraction

To understand the **index of refraction** (often called the **n-value**), imagine light as a champion sprinter. In the empty void of outer space, light travels at its absolute top speed, about 186,000 miles every single second. (299,792,458 meters per second) Nothing in the universe moves faster.

However, when light hits a transparent material like a glass window, it doesn't just pass through at full speed. It hits a "traffic jam" of atoms. The light interacts with the particles inside the material, which causes it to drag and slow down. The **n-value** is just a simple number that tells us exactly how much a material "brakes" or slows down light (photons) (Hewitt, 2014).

The Mathematical Speed Limit

Scientists calculate this value by comparing the speed of light in a vacuum to its speed inside the material.

Air (n = 1.00): Light travels (for practicable purposes) almost as fast in air as it does in space.

Water (n = 1.33): Light slows down by about 25%.

Glass (n = 1.50 to 1.90): Depending on the type of glass, light can be forced to "crawl" at nearly half its original speed.

When light slows down as it enters a material at an angle, it doesn't just decelerate; it changes direction. Imagine a lawnmower moving from a flat sidewalk into thick, tall grass at an angle. If the right wheel hits the grass first, it slows down while the left wheel is still on the fast sidewalk. This lopsided change in speed causes the whole lawnmower to pivot and turn (Walker, 2014).

Light behaves exactly the same way. This bending is called **refraction**. The higher the n-value, the more the material bends the light toward a new direction.

N-Value
Index of Refraction

$n = c / v$

- n = Index of Refraction
- c = speed of light in a vacuum (299,792,458 \meters/sec)
  OR
- c = speed of light in a vacuum (186,000 miles/sec
- v = speed of light in a transparent material

Typical n values

- Vacuum = 1.00
- Water = 1.33
- Plastic Lenses (CR-39) = 1.498
- Spectacle Crown Glass = 1.523
- Trivex = 1.53
- Polycarbonate = 1.586
- Flint Glass = 1.61
- Hi-Index Plastic = 1.67 or 1.74

N-Value cont.

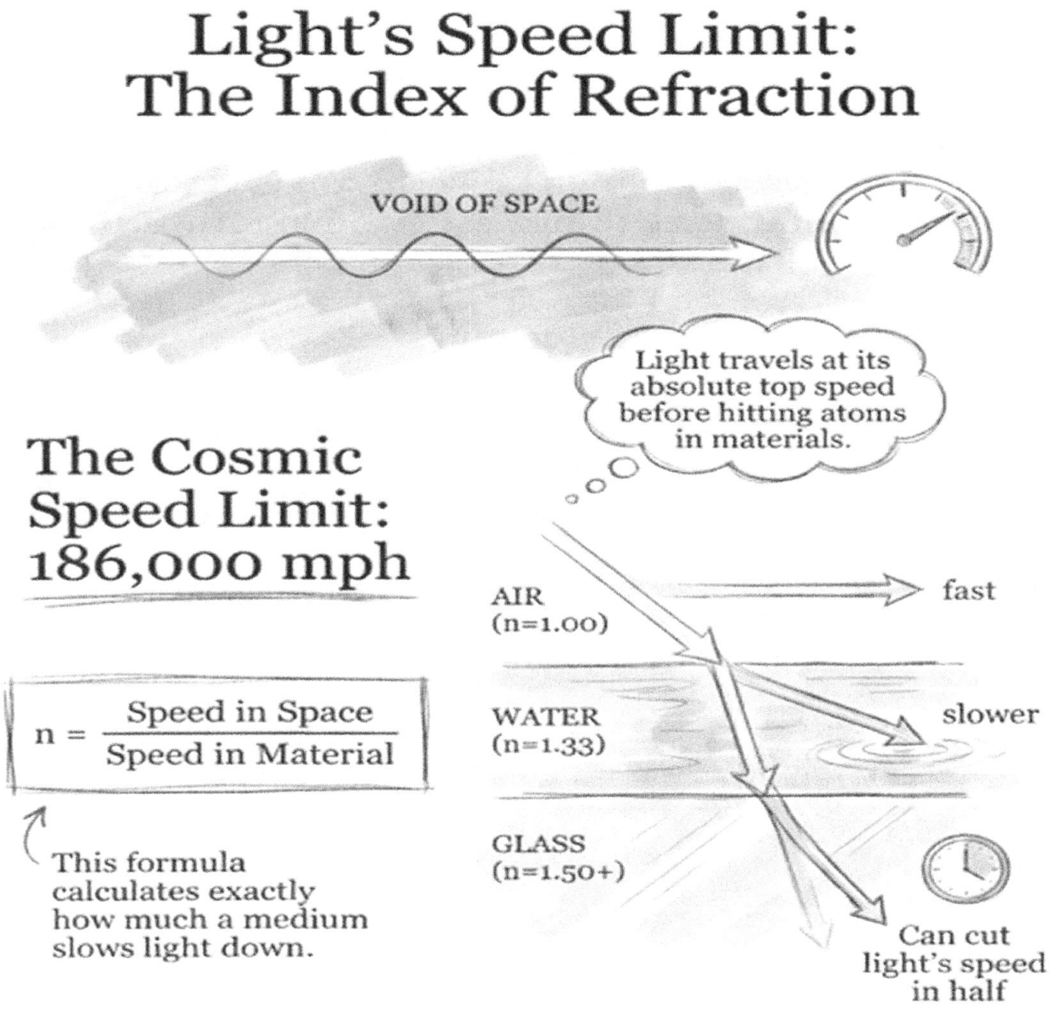

N –value is indicative of the **density** of the material. The denser the material, the more light is slowed down and the greater it will **refract**. Greater refraction allows **flatter curves** resulting in a **thinner lens** compared to a lower n-value material in a similar prescription.

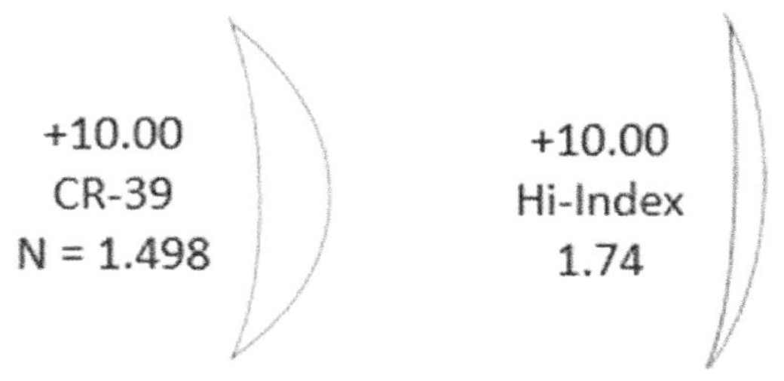

Lens Color Dispersion

Abbe Value

Abbe Value assigns a number to a lens material that indicates how much color break-up (or Rainbow Effect) is inherent in that material. It is generally of little concern but can cause issues if a high prescription is coupled with a very large frame, something best avoided for weight and peripheral distortion reasons.

The higher the Abbe Value, the LESS dispersion or Rainbow Effect.

Lens Material	Abbe Value
Polycarbonate	30
1.67 Plastic	32
1.56 Mid-Index Plastic	36
1.70 Hi-Index Glass	39
Trivex	43
Plastic CR-39	58
1.523 Crown Glass	59

MORE Color Dispersion

LESS Color Dispersion

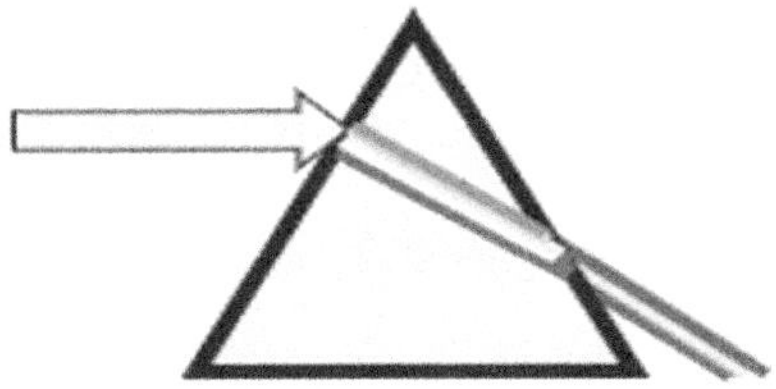

Imagine looking through glass and seeing shimmering rainbow halos dancing around the edges of every single object you try to focus on. This colorful blur happens because different colors of light travel through lenses at slightly different speeds (Hewitt, 2014). Light possesses hidden chaos. The Abbe value identifies how much a lens material "scatters" these colors into a messy, distracting rainbow (Walker, 2014).

High Abbe values indicate that **a material is excellent** at **keeping light together**, ensuring your view remains crisp, white, and sharp. **Low values** cause distracting **color fringes** that frustrate the human eye. Today's lenses are designed to minimize this effect. It is rarely an issue absent extremely strong prescriptions but you need to understand the concept for certification.

Lens Materials Summary

Lens Material	Abbe	N Value	Notes
Polycarbonate	30	1.586	Most common material today, most impact resistant, blocks UV
Plastic CR-39	58	1.498	Optics better than Poly, less impact resistant, thicker than poly
Hi-Index Plastic	32 to 36	1.60 to 1.74	Good option for high scripts, thinner, but more expensive
Trivex	45	1.54	Premium poly, better optics, thicker in higher scripts, very impact resistant, blocks UV
Crown Glass	59	1.523	Widely used until 1980's, less impact resistance, good optics, very scratch resistant, heavy
Hi-lite Glass	31	1.70	Rarely used, poor impact resistance, thinnest option with high scratch resistance but heavy

Chapter Eleven

Lens Designs

OPHTHALMIC lens design has transitioned from rudimentary spherical glass to computationally optimized surfaces that compensate for complex optical aberrations. Each design represents a historical milestone in the attempt to reconcile human physiology with the laws of physics (Walker, 2014). Innovation follows the curve. Geometry dictates the resolution.

Single Vision (SV) Lenses

Single vision remains the most fundamental design, featuring a single dioptric power across the entire lens surface.

History: The earliest recorded use of corrective lenses dates to the 13th century in Italy, primarily as "reading stones" made of beryl or glass. By the late 1800s, mass production of meniscus (curved) lenses replaced flat or "periscopic" designs to reduce peripheral distortion (Hewitt, 2014).

Function: They correct for one focal distance, typically distance, intermediate, or near, making them the primary choice for patients with myopia, hyperopia, or astigmatism **who do not yet have presbyopia**. The non-presbyopic patients can see both far and near with a single vision lens because they still have plenty of "**accommodation**" for near work.

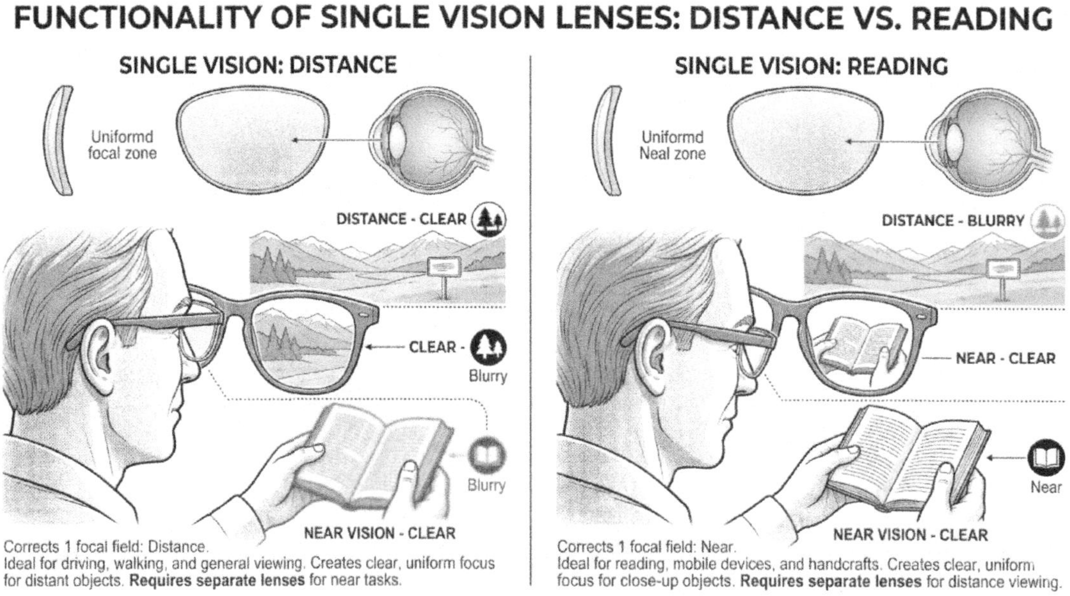

Multifocal Lenses: Bifocals and Trifocals

These designs utilize physical segments to provide multiple focal points within a single frame.

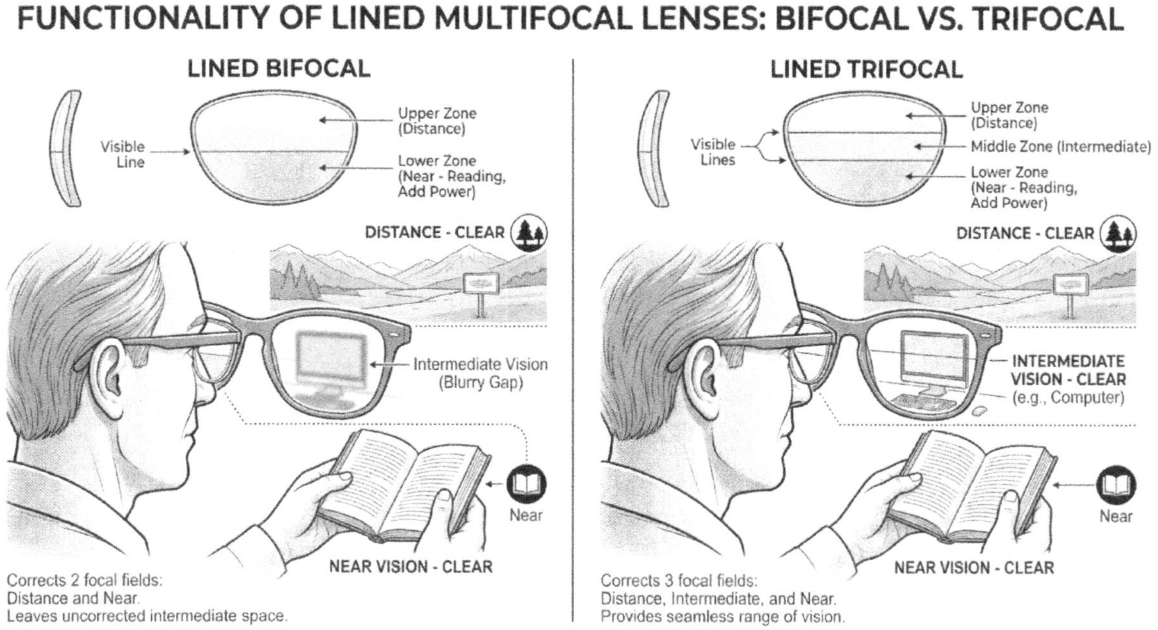

Bifocals (Flat-top or D-seg): Concept was invented by Benjamin Franklin in 1784 by cutting two different lenses in half and joining them in a single frame. In the early 20th century, the fused "Flat-top" became the industry standard, allowing the wearer to look down through a "D-shaped" segment for near vision (Serway & Jewett, 2018).

Trifocals: Introduced in the mid-1940s to address the "intermediate gap" caused by advancing presbyopia. They add a third horizontal segment above the near portion, calibrated specifically for tasks at arm's length, such as viewing a dashboard or sheet music.

Progressive Addition Lenses (PALs)

Progressive lenses provide a seamless transition of power from the top (distance) to the bottom (near) without visible lines.

History: Although patents existed as early as 1907, the first commercially successful progressive, "Varilux," was launched in 1959 by Bernard Maitenaz. Early designs were plagued by narrow "corridors" of clear vision and significant "swim effect" (peripheral distortion) (Walker, 2014).

Function: Modern PALs utilize complex aspheric curves to provide a natural viewing experience at all distances.

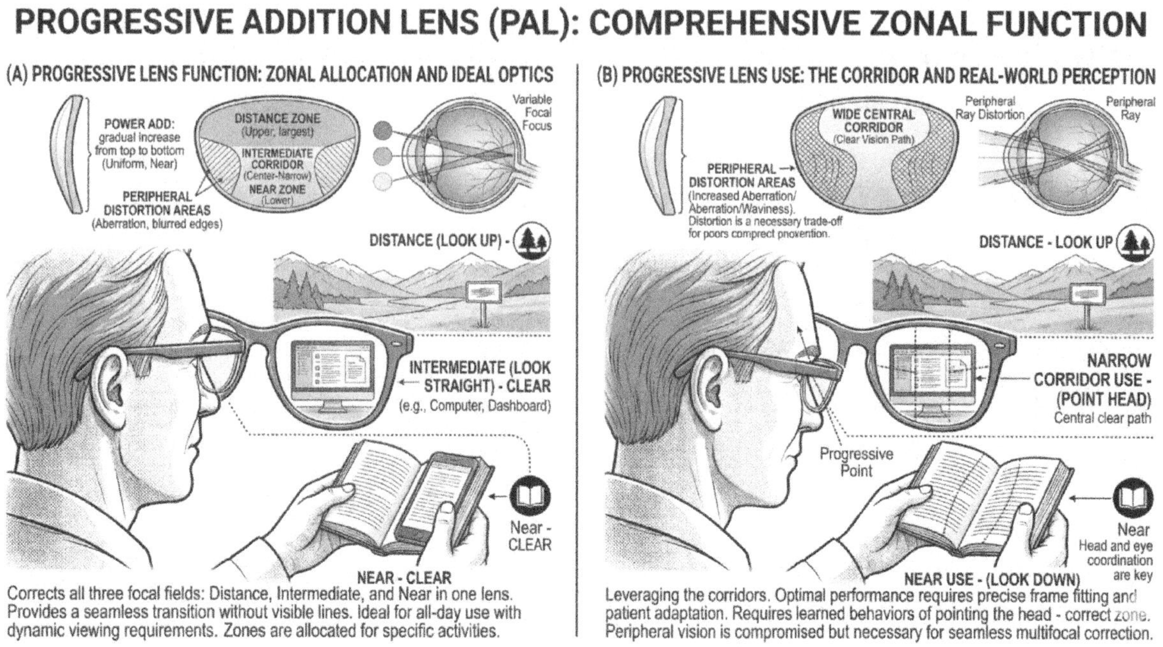

Aspheric and Atoric Designs

Aspheric lenses deviate from the standard spherical shape, flattening toward the edges.

History: Developed in the late 20th century, asphericity was initially used in high-power "cataract eyeglass lenses" (before surgical implants) to reduce weight and thickness.

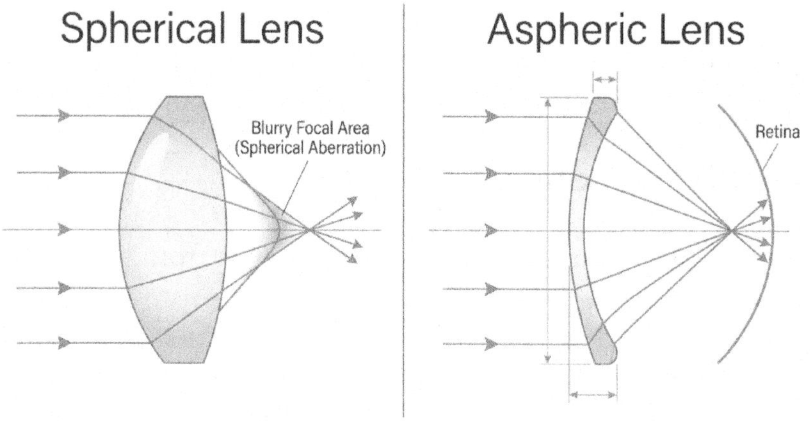

Function: By gradually changing the curvature from the center to the periphery, these lenses eliminate "spherical aberration" and provide a thinner profile with less "eye magnification" or "minification" for the wearer (Serway & Jewett, 2018).

Free-form (Digital/HD) Design

Free-form is not a material but a manufacturing process that uses Three-Axis CNC generators to surface lenses with point-by-point precision.

History: Emerging in the early 2000s, digital surfacing allowed manufacturers to move beyond "static" molds.

Function: This design allows for "High-Definition" optics tailored to the specific frame dimensions and the wearer's unique vertex distance and "wrap" angle. It effectively eliminates the peripheral blur found in traditional progressive designs. Proper and accurate measurements are essential.

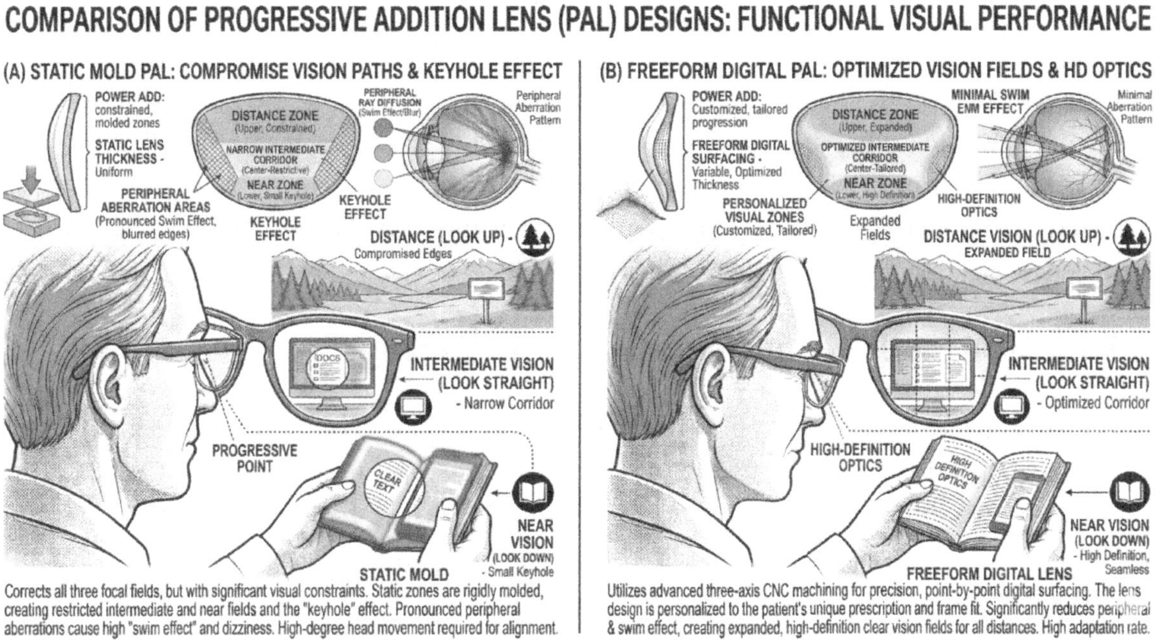

The evolution of ophthalmic lens design represents a transition from rudimentary, static geometry to complex, digitally-optimized surfaces that mimic the eye's natural focus. This progression mirrors the history of material science and mathematical modeling (Hewitt, 2014; Walker, 2014).

The Era of Discrete Segments

Initial lens designs were exclusively Single Vision, featuring a uniform spherical curvature across the entire lens to correct a single refractive error. This sufficed until the onset of presbyopia, the age-related stiffening of the eye's internal lens.

Later, the Trifocal emerged to address the "intermediate gap", the blurry zone between the horizon and a book. These designs utilized visible, raised segments to partition functional areas. While functionally sound, these "lined" lenses created a harsh "image jump" and a blur line as the pupil crossed the physical boundary between different optical powers (Walker, 2014).

The Shift to Continuous Gradients

The mid-20th century introduced Progressive Addition Lenses (PALs), which eliminated visible lines in favor of a seamless vertical corridor of increasing power. This design utilized Aspheric geometry, where the surface curve flattens toward the edges to reduce thickness and peripheral distortion (Hewitt, 2014). Early versions were "Static Molds," meaning the power transition was cast into the plastic at a factory using a pre-set template. These "hard" designs created considerable blur on the edges forcing a pronounce "point the nose" effect.

Static designs were rigid. One size rarely fit all. These lenses often suffered from a "keyhole effect," where the clear viewing area was restricted to that narrow central path, surrounded by a "swim effect" of blurry peripheral vision that could make a wearer feel dizzy (Serway & Jewett, 2018).That was often the cause of patient rejection of the lens style and considerable chair time troubleshooting and coaching.

The Digital Revolution: Free-form Surfacing

The current pinnacle of design is Free-form Digital Surfacing. This is not a specific shape but a manufacturing philosophy that uses three-axis CNC generators to generate the lens surface with point-by-point precision (Walker, 2014). This allows for "High-Definition" optics tailored to the wearer's specific frame tilt, vertex distance, and wrap angle.

Algorithms optimize the view and the software greatly reduces the edge distortions. Modern digital designs have effectively widened the clear corridors and minimized peripheral aberrations, providing a visual experience that is closer to natural, youthful sight than ever before (Hewitt, 2014).

(PALs incorporate nearly invisible markings that serve as a permanent "biological record" of the lens's design and orientation. These markings are essential for both the initial verification of the prescription and the identification of the lens type years after the initial purchase (Walker, 2014). The data is permanently etched.

The Role of Laser Etchings

Every modern PAL is engraved with microscopic laser markings, typically located on the nasal and temporal sides of the lens (Hewitt, 2014). These etchings usually include:

The Manufacturer's Logo: A unique symbol identifying the brand and specific design model.

The "Add" Power: Two or three digits (e.g., "25" for a +2.50 add) indicating the reading strength.

Alignment Circles: Two small circles precisely 34 mm apart that define the horizontal axis of the lens. Because these markings are permanent, an optician can use them to "re-mark" a lens even if the original ink stamps have been cleaned away (Serway & Jewett, 2018). This process ensures the optical center remains can be aligned with the wearer's pupil during initial and subsequent dispenser adjustments.

.**The PAL Identifier Chart:** Since there are hundreds of progressive designs on the market, professionals use a PAL Identifier Chart to decode these laser symbols. This comprehensive database acts as a "Rosetta Stone" for the optical industry. By matching the etched logo on the lens to the entry in the chart, the optician can determine exactly which technology is being used and verify the minimum fitting height required for the frame (Walker, 2014).

Verification and Alignment: During a quality check, an optician places the lens over a physical layout chart. They align the laser-etched circles with the markings on the chart to "re-ink" the lens with temporary stamps (Hewitt, 2014). These stamps show exactly where the distance zone, intermediate corridor, and reading area are located. If the alignment circles are tilted or off-center, the wearer will experience the "swim effect" or narrower fields of view.

Verification protects the patient and the dispenser. Without these hidden markers, adjusting or troubleshooting a progressive lens would be purely guesswork.

Layout Chart Data

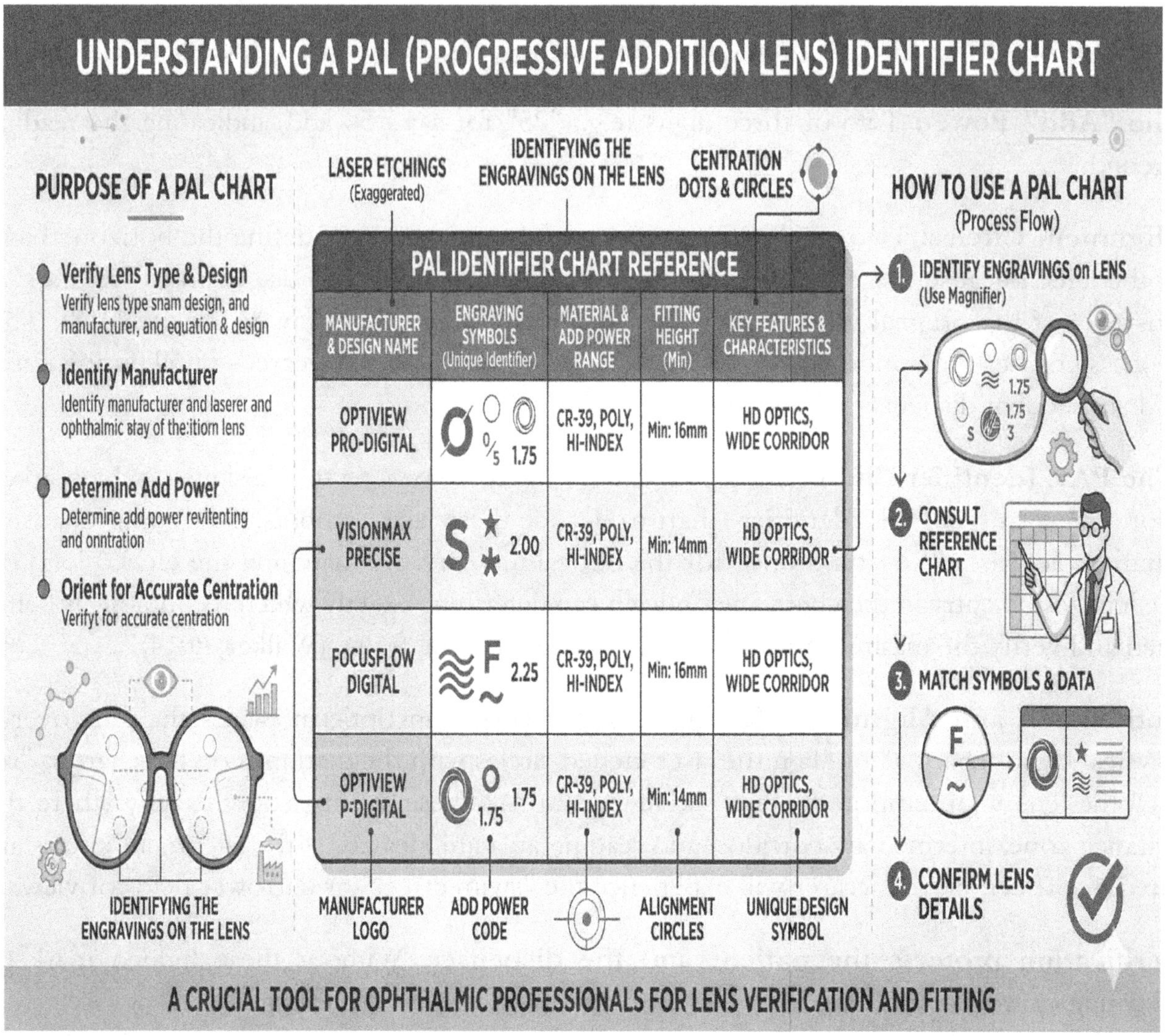

Chapter Twelve

Lens Options

Anti-Reflective Properties

THE purpose of anti-reflective (AR) coatings is to maximize the amount of light that actually passes through an ophthalmic lens while simultaneously eliminating the distracting ghost images and glare that bounce off the surface (Hewitt, 2014). For a standard plastic lens, approximately 8% to 12% of incoming light never reaches the eye because it is reflected away. This loss of transmission creates a dimmer view for the wearer. AR technology minimizes this by utilizing the physics of wave interference to cancel out these unwanted reflections.

The Mechanism: Destructive Interference

AR coatings are composed of multiple microscopic layers of metallic oxides. These layers are vacuum-coated onto the lens surface in a clean-room environment (Walker, 2014). To function, each layer must be engineered to a precise thickness, exactly one-quarter of a wavelength of light. This allows the coatings to utilize a phenomenon called destructive interference.

When a light wave hits the front of the AR layer, some of it reflects back immediately. The rest travels through the layer, hits the lens surface, and then reflects back. Because the layer is exactly a quarter-wavelength thick, the second reflection is "out of phase" with the first one by half a wavelength. The peaks of the second wave line up perfectly with the valleys of the first. They cancel each other out. The reflection essentially disappears because the waves destroy each other by applying physics and geometry.

The Benefit: Increased Transmission

By "destroying" the reflections, the law of conservation of energy forces the light to travel *through* the lens instead of reflecting away (Serway & Jewett, 2018). This increases light transmission to nearly 99%.

This provides three primary advantages for patients:

Night Vision: It reduces the "starburst" effect around headlights and streetlamps by removing internal lens reflections (Walker, 2014).

Cosmetic Clarity: It makes the lenses appear less visible to others, allowing people to see your eyes rather than a mirror-like glare.

Digital Comfort: It reduces the distracting glare from computer screens and overhead fluorescent lighting, which helps minimize eye fatigue during long study sessions (Hewitt, 2014).

Modern AR coatings often include additional hydrophobic (water repelling) and oleophobic (oil repelling) top layers that discourage fingerprinting and smudging compared to older products, assisting the optical surface to remain clear between cleanings.

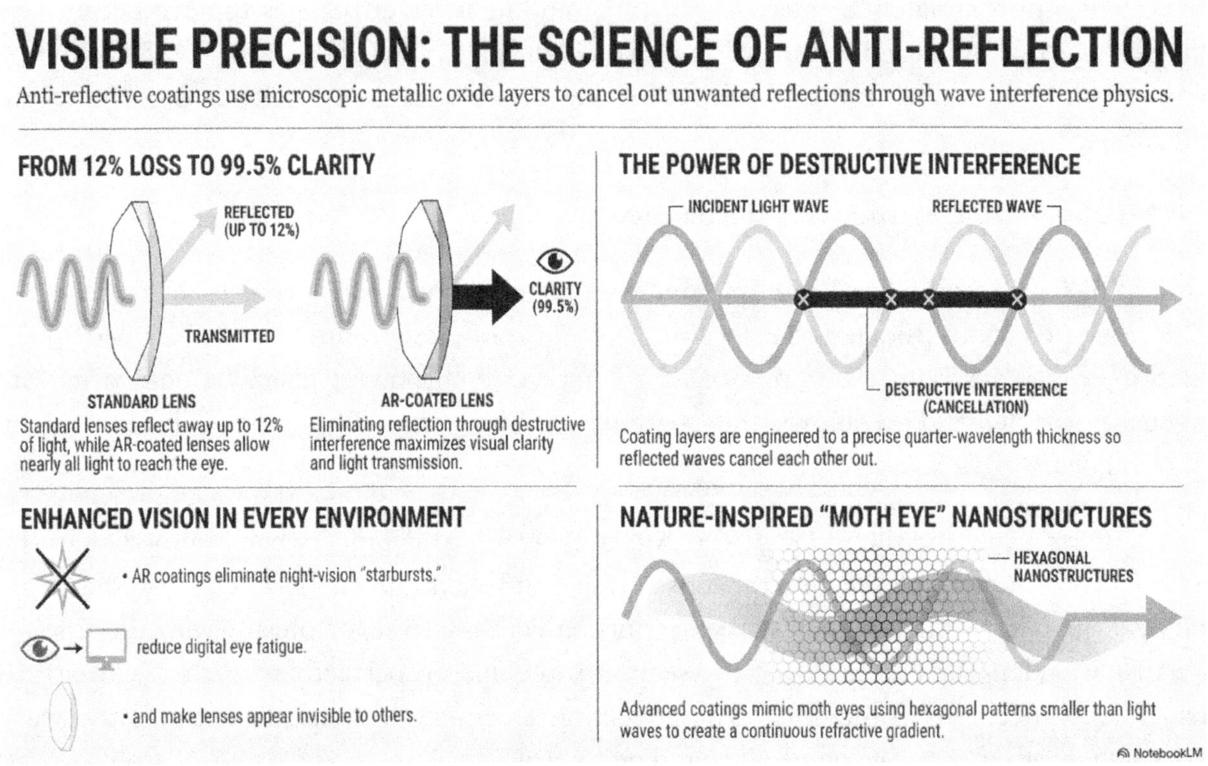

Blue Light Blockers

"Blue light blockers" refers to ophthalmic lens technologies designed to selectively filter or reduce high-energy visible (HEV) light, specifically in the 400nm to 450nm wavelength range (Livingstone, 2014). This segment of the visible spectrum is adjacent to ultraviolet (UV) radiation and possesses sufficient energy to penetrate the ocular media and reach the retina. Filtering provides a shield.

The Science of Blue Light Blocking Manufacture

Substrate Absorption and **Reflective Interference**

Substrate Absorption (In-Mass): During the polymer casting process, specific chemical "monomers" or pigments are infused directly into the lens material. These molecules are geometrically tuned to absorb HEV wavelengths while allowing longer lower-energy wavelengths (like green and red) to pass through. This often results in a subtle yellow or "champagne" tint, as the removal of blue light leaves the warmer colors of the spectrum dominant (Hewitt, 2014).

Reflective Interference (Surface Coating): This method uses the same "quarter-wavelength" physics as anti-reflective coatings. Instead of maximizing transmission, engineers apply specialized metallic oxide layers that are designed to reflect a specific percentage of blue light away from the eye (Serway & Jewett, 2018). These lenses typically exhibit a distinct blue or purple residual reflection on the front surface.

Practical Use and the Science

The medical community distinguishes between "digital eye strain" and permanent "retinal damage." Confirmed science primarily supports the following practical applications:

1. Circadian Rhythm Regulation

Scientific consensus confirms that exposure to short-wavelength blue light suppresses the secretion of melatonin, the hormone responsible for sleep-wake cycles (Wolfe et al., 2020). Using blue blockers in the evening helps prevent the "alertness signal" that occurs when light hits the intrinsically photosensitive retinal ganglion cells (ipRGCs). Many digital devices now have settings to reduce HEV in the evening.

2. Visual Contrast and Comfort

In physics, short-wave blue light scatters more easily than long-wave light, a phenomenon known as Rayleigh Scattering (Hewitt, 2014). This scattering creates "chromatic noise" or a hazy glare on digital screens. By filtering these fuzzy wavelengths, blue blockers can increase perceived contrast and reduce the effort required for the eye to focus during prolonged computer use.

3. The Retinal Damage Debate

Currently,(2025) there is no verified clinical evidence in humans suggesting that the amount of blue light emitted by digital devices is sufficient to cause age-related macular degeneration (AMD) or physical ocular injury (Livingstone, 2014). The sun remains the primary source of dangerous HEV light. Protection is precautionary.

Ultra Violet Light Blocking

Ultraviolet (UV) blocking in ophthalmic lenses is a critical safety feature designed to shield the ocular structures from radiation ranging from 100nm 400nm. Unlike visible light, UV photons possess enough energy to break chemical/biological bonds, leading to cumulative biological damage (Hewitt, 2014).

Inherent Material Properties

Different lens materials possess varying natural abilities to absorb UV radiation based on their molecular density and chemical composition (Walker, 2014).

- **CR-39 (Standard Plastic):** This material is a "loose" polymer. It naturally allows approximately 10% to 15% of UVA light to pass through. It requires a supplemental chemical bath, a "UV dip" to reach near 100% protection. It should be recommended with CR-39.
- **Polycarbonate and Trivex:** These are "high-index" materials with a tightly packed molecular lattice. They are inherently UV-opaque up to 380nm to 400 nm without any extra treatments (Serway & Jewett, 2018). UV treatments are not needed.
- **Glass:** Standard crown glass is a poor UV blocker. It transmits significant amounts of high-energy radiation unless it is specially tinted or treated (photchromatics) during the melting process.

The Manufacturing Science

To achieve "UV 400" protection in materials that aren't naturally opaque, manufacturers use two primary techniques (Walker, 2014):

In-Mass Additives: During the liquid monomer stage, UV-absorbing chemicals (benzotriazoles) are mixed directly into the material. These molecules act as "energy sponges," capturing UV photons and converting their energy into harmless heat before they pass through the lens (Hewitt, 2014).

Thermal Dipping: Finished lenses are submerged in a heated solution of UV-inhibitors. The heat expands the polymer's surface pores, allowing the protective chemicals to migrate into the front few microns of the lens.

Practical Use and Confirmed Science

The medical necessity of UV blocking is grounded in established ocular pathology. Science confirms that excessive exposure to UV radiation is a direct catalyst for several conditions (Goldstein & Cacciamani, 2022):

Photokeratitis: Essentially a "sunburn" of the cornea (like staring at an eclipse). *High-intensity* UV exposure causes the death of epithelial cells, leading to temporary but extreme pain (Wolfe et al., 2020).

Cataractogenesis: UV-B radiation is a primary driver of protein "clumping" within the eye's crystalline lens. This clouding assists in developing cataracts, which is a leading cause of blindness globally (Livingstone, 2014).

Pterygium: UV exposure triggers the growth of "surfer's eye," a fleshy tissue that can grow across the cornea and obstruct vision in severe cases.

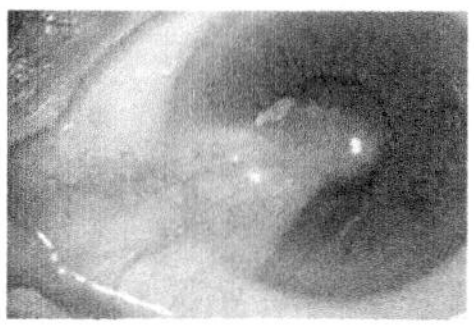

WIKI

Confirmed data shows that UV protection is not an "option" so much as a requirement for long-term ocular health.

Photochromic Technology

Photochromic technology allows ophthalmic lenses to function as a "dynamic filter" that adjusts its transmission of visible light based on the intensity of ambient ultraviolet (UV) radiation. This process is essentially a reversible chemical reaction that occurs within the lens substrate or on its surface (Hewitt, 2014). Transitions is a specific brand name for the modern plastic version of this technology. Chemistry manages the light.

The Mechanism: Molecular Transformation

The science of darkening differs significantly between traditional glass lenses and modern plastic (Transitions) designs.

Photochromic Glass: These lenses contain microscopic crystals of silver halide (usually silver chloride) embedded throughout the glass during the melting stage (Walker, 2014). When UV photons strike these crystals, they trigger a reaction where the silver ions gain an electron and transform into clusters of metallic silver. These metallic clusters are opaque and scatter visible light, causing the lens to darken (Serway & Jewett, 2018).

Plastic (Transitions) Lenses: Because plastic cannot withstand the heat required to suspend silver crystals, modern lenses use organic molecules (Hewitt, 2014). In their resting state, these molecules are folded and transparent. When exposed to UV, the chemical bonds "snap" open into a flat, planar structure that absorbs visible light. As the UV source is removed, the molecules naturally fold back into their clear state.

The Science of the Manufacturing Process

Transitions and other plastic photochromics utilize a process called **imbibition** or **trans-bonding** (Walker, 2014). Rather than being mixed into the liquid plastic, the photochromic molecules are applied to the front surface of the finished lens. The lens is heated in a specialized chamber, which "opens" the surface pores and allows the molecules to sink approximately 150 to 200 microns deep into the material. This ensures that the darkening is uniform across the lens, regardless of its thickness. Surface integration ensures consistency.

UV and Visible Light Blocking Abilities

The performance of a photochromic lens is influenced by its material composition and the surrounding environment (Goldstein & Cacciamani, 2022).

Inherent UV Protection: Regardless of whether they are clear or dark, almost all modern Transitions lenses (especially those made of Polycarbonate or High-Index materials) block 100% of UVA and UVB radiation (Livingstone, 2014). The chemistry that causes darkening *requires* UV energy to function, meaning the lens acts as a "UV sponge" at all times.

The Temperature Paradox: Because the "fading" process is triggered by thermal energy, photochromic lenses perform better in cold weather. In high heat, the molecules are constantly trying to fold back into their clear state even as UV tries to open them (Walker, 2014). This results in the lens not getting as dark on a hot summer day, as it would on a crisp, snowy winter morning.

The Windshield Constraint: Standard photochromic lenses do not darken inside a car because modern automotive windshields are manufactured with a UV-absorbing interlayer. Without UV "fuel," the molecules remain folded and clear. Specialized designs exist that utilize the lower-energy visible light spectrum to darken behind glass (Xtra-Active).

Polarization

Polarization options for ophthalmic lenses are designed to selectively filter out blinding glare reflected from horizontal surfaces like water, snow, and wet roads. While standard sunglasses simply reduce overall brightness, polarized lenses solve a specific physics problem to improve visual comfort and clarity.

The Rope through the Fence:

To understand how polarized lenses work, imagine unpolarized light from the sun as a rope oscillating randomly in many different planes, up/down, side-to-side, and everywhere in between. Light reflected off a flat surface becomes "organized" (polarized), oscillating primarily in one horizontal plane.

If you feed the rope through a slot between the pickets of a vertical fence, only the waves shaken up and down (vertically) pass through. If you shake the rope side-to-side (horizontally), the fence pickets block the waves. A polarized lens functions as a microscopic vertical fence for light waves.

The Manufacturing Science: Stretching and Doping

The key component of a polarized lens is a microscopically thin filter film made of Polyvinyl Alcohol (PVA). The manufacturing process relies on molecular alignment.

Heating and Stretching: The initial PVA film has molecules arranged in a messy, random tangle. To organize them, the film is heated and stretched to five times its original length. This extreme tension forces the long-chain polymer molecules to unfold and align parallel to each other, creating a directional structure (Vision Ease, 2016).

Doping (Dyeing): This stretched, aligned film is delicate and transparent. It is dipped into a solution of iodine or dyes. These molecules absorb into the aligned polymer chains, adhering along their length like dark ink. These aligned chains of "doped" molecules create an invisible microscopic grid of parallel conducting “fence pickets” (Reddit, 2014).

The Mechanism of Blocking Visible Light

The aligned microscopic "wires" in the polarizing film interact exclusively with light waves based on their orientation.

In unpolarized light, half of the waves have a horizontal orientation, and half have a vertical orientation. A standard clear lens transmits nearly 100% of this light. A polarizing filter, simply by blocking one plane, immediately reduces unpolarized light transmission by approximately 50% (ZEISS Group, 2025).

In a finished ophthalmic lens, this filter film is oriented **vertically** within the material (the "microscopic fence wires" run up-and-down) (Vision Ease, 2016).

When horizontally polarized reflected glare (side-to-side waves) hits the vertically oriented lens, the side-to-side waves are absorbed by the molecular grid, effectively "erasing" the glare from the wearer's view. Waves that possess a vertical component are transmitted, allowing you to see objects clearly without the blinding shine (Reddit, 2014). This results in a lens that blocks most horizontal glare while still allowing plenty of useful light to pass through. **An easy way to check polarization is to look at any reflected glare (like off a shiny floor surface) and turn the lens 90 degrees. If the glare changes you have polarization!**

Mirror Coatings

Mirror coatings on ophthalmic lenses serve a dual purpose: they provide a bold, reflective aesthetic while significantly enhancing visual comfort by reducing the total amount of light that reaches the wearer's eyes. Commonly applied to sunglasses (especially polarized ones), these coatings are engineered using sophisticated thin-film physics to "bounce" specific colors away before they can transmit through the lens (Hewitt, 2014; Walker, 2014).

The Physics of Shine: Reflecting Visible Light

Unlike a standard lens that aims to maximize light transmission, a mirrored lens is designed to obstruct it. Anti-reflective coatings use **destructive interference** to make waves disappear. A mirror coating uses **constructive interference** to make them stronger. The lens surface is covered with microscopic layers of specific metallic or non-conducting (dielectric) materials (Hewitt, 2014).

When light hits the mirror, some reflects off the front of the top layer, while some passes through and reflects off the second layer. To create a strong mirror effect, engineers precisely tune the thickness of these layers (often based on 1/2 the wavelength of the target color). This ensures that the reflected waves from both layers align perfectly, peak-to-peak and valley-to-valley. They "add up," creating an intense, bright reflection of that specific color, while the other colors pass through the lens (Walker, 2014).

Vacuum and Vapor: The Manufacturing Science

Creating an optical mirror requires extreme precision, as the layers must be flawless and measured in nanometers. This cannot be "painted" on; it must be grown through a process called vacuum deposition (sometimes called sputtering) (Vision Ease, 2016).

The Vacuum Chamber: The lenses are loaded into a specialized, high-vacuum chamber, similar to deep space. All air is removed to prevent contamination during the coating process.

Vaporization: Inside the chamber, target materials—such as titanium dioxide (which creates brilliant colors) or silica, are bombarded with an electron beam. This intense energy vaporizes the solid material into a fine, energetic cloud of molecules (Walker, 2014).

Condensation: This vaporized cloud travels in a straight line (since there is no air to block it) and condenses uniformly onto the surface of the spinning lenses. The machine applies multiple distinct layers, alternating materials and thicknesses point-by-point to create the final designed mirror effect (Serway & Jewett, 2018).

Flash vs. Full: Mirror Options

Optical laboratories offer several mirror designs based on how much light needs to be reflected:

Full (Solid) Mirrors: These are highly efficient, solid-color mirrors (like intense silver, gold, or blue) that maximize light reflection. They often look completely opaque from the outside. A full mirror can reduce the amount of light reaching the eye by an *additional* 10% to 15% beyond what the dark base tint provides (ZEISS Group, 2025). They are ideal for extremely high-glare environments like open water or snowfields.

Flash (Gradient) Mirrors: These are much subtler, creating only a slight hint of color and reflection. A "flash" mirror might only provide a 3% to 5% boost in light reduction (Vision Ease, 2016). They provide a modern cosmetic "pop" without making the view significantly darker for the wearer.

Colors and Interaction: While the mirror *color* (blue, red, green) is primarily cosmetic, it interacts with the base lens tint. For example, a bright red mirror reflects high-energy red light away. This makes the base gray tint look slightly cool (more bluish) to the wearer because the warm red light is missing (Hewitt, 2014).

Tinted Lenses

Ophthalmic lens tinting is a dynamic field that combines chemistry, material science, and thin-film physics to enhance visual comfort, provide specialized performance, or offer aesthetic customization. The core purpose of any tint is not to "add color," but rather to act as a precise optical filter that reduces or absorbs specific bands of incoming visible light (Hewitt, 2014; Walker, 2014).

The Mechanism: Selective Absorption and Dimming

To understand how tints work, we must recognize that visible sunlight is composed of all the colors of the rainbow, each traveling at a different wavelength and carrying a specific energy (approx. 400 nm to 700nm).

A tint is essentially a chemical "dye" composed of specialized molecules known as chromophores (colored chemical compound). These molecules possess a specific geometric structure that makes them resonate with the incoming photons of precise wavelengths (Walker, 2014).

Blocking Specific Colors: When a photon (say, a high-energy blue light photon) strikes a chromophore that is "tuned" to its energy, the molecule absorbs the photon's energy. This excites the molecule, often causing an electron to jump to a higher energy level (excitation) (Serway & Jewett, 2018). Instead of passing through, that light energy is trapped by the dye and converted into imperceptible heat within the lens. The law of conservation of energy ensures that this subtracted light never reaches the eye.

Resulting Color Perception: A tinted lens appears a certain color (e.g., yellow) because the dye has absorbed the conflicting colors (blue and violet) and *allowed* the yellow and red wavelengths to transmit (pass through) relatively uninhibited (Hewitt, 2014). Gray tints are the optical standard because they absorb all colors relatively evenly across the spectrum, just dimming the entire world without distorting natural color perception (Walker, 2014).

Vacuum and Dip: The Manufacturing Science

The manufacturing process for tinting depends entirely on whether the lens material is traditional **glass** or modern **plastic**.

Plastic (CR-39 or Polycarbonate) Lenses: Imbibition (Dip Dyeing)

This is the most common custom method for ophthalmic labs, as plastic (polymer) lenses are fundamentally porous at a molecular level, especially when heated.

The Process: Finished lenses are placed into a mechanical arm and submerged into heated water baths (approx. 90°{C} 195°{F}) that are saturated with organic dyes.

The Interaction: The intense heat causes the tightly wound polymer chains of the plastic lens to expand and "open up." In a process called imbibition (molecular absorption), the tiny dye molecules migrate into these expanded molecular pores (Walker, 2014). The longer the lens stays in the bath, the deeper the dye penetrates and the darker the tint becomes. Once removed and cooled, the plastic substrate contracts, permanently locking the dye molecules inside the surface layers. (Reddit, 2014).

Glass Lenses: In-Mass (Fused) Chemistry

Glass is rigid and non-porous; you cannot dip-dye it. Tints for glass lenses must be created during the initial manufacturing stage by fusing metal oxides directly into the molten glass batch *before* it is cooled into a lens blank (Walker, 2014).

For example, adding *cob*alt oxide results in blue glass, while adding nickel or iron oxides can create gray or green shades (Serway & Jewett, 2018). This ensures the color is uniform throughout the entire lens structure, not just on the surface.

Application Options: Solid vs. Gradient vs. Sun Tinting

Optical laboratories customize tints into different application categories based on user needs:

Solid Tints

A solid tint provides perfectly uniform density (darkness) across the entire lens surface. To manufacture this, the lens is held in a mechanical clamp and lowered into the dye bath completely horizontally (or rotated slightly) for a precise, continuous period of time. *Use:* Standard sunglass dimming or specialized performance (e.g., skiing, hunting) (ZEISS Group, 2025).

Gradient Tints

A gradient tint is darker at the top and progressively lighter toward the bottom. This effect cannot be created by a single dip. Instead, the lab uses a mechanical arm that slowly and precisely lifts or lowers the lens into the dye bath over several distinct "stages." The top section of the lens spends the *most* total time submerged (becoming darkest), while the bottom section spends the *least* time (remaining light or clear). *Use:* High fashion, driving (dimming the sky while dashboard is clear), or reading outdoors.

Sun Tinting

This isn't a technique but a performance category defined by VLT (Visible Light Transmission).

A dark sunglass tint typically has a VLT of 15% to 25%, meaning it successfully subtracts 75% to 85% of incoming visible light.

It is important for to understand that a dark tint *must* be combined with 100% UV protection. When a lens is darkly tinted, it forces the pupil to dilate (widen), just like a camera aperture in low light. If the lens lacks proper UV blocking, this dilated pupil will inadvertently allow far more dangerous UV radiation to strike the retina than if no glasses were worn at all, accelerating long-term damage (Hewitt, 2014).

Scratch Resistance

Scratch-resistant (SR) coatings, often referred to as "hard coats" in the optical industry, are optically clear, microscopic layers applied to the surfaces of ophthalmic lenses. Their primary purpose is to enhance the surface hardness of modern plastic lens materials, which are inherently soft and vulnerable to abrasion during daily use and cleaning (Hewitt, 2014; Walker, 2014). It is critical to understand that these coatings are scratch-resistant, **not scratch-proof.** While they significantly extend the life of the lens, they can still be damaged by excessive force or improper handling or cleaning.

The Material Science: The Hybrid Matrix

To understand how an SR coating functions, we must examine its unique molecular composition. The goal is to create a layer that is incredibly hard, yet flexible enough not to crack when the underlying lens expands or contracts due to temperature changes.

SR coatings are typically **hybrid materials** known as polysiloxanes (Walker, 2014).

The Structure: Imagine a dense microscopic "chain-link fence" (a polymer matrix) made of organic molecules. Embedded within this flexible fence are countless tiny particles of inorganic material, usually silica (silicon dioxide, the primary component of glass) (Serway & Jewett, 2018).

The Interaction: The silica particles provide the extreme surface hardness, creating a barrier that resists being "plowed" by abrasive objects. The surrounding polymer matrix provides the necessary flexibility (elasticity), allowing the coating to absorb subtle impacts and thermal stress without shattering or delaminating (peeling off). This dense, cross-linked molecular network is what provides the durability.

The Manufacturing Science: Application and Curing

The application of a hard coat is a high-precision manufacturing step that occurs before any anti-reflective (AR) treatments. There are two primary methods:

Dip Coating (Thermal Cure)

This is the most common method for bulk manufacturing or prescription laboratories.

Process: Finished lenses are held vertically in a mechanical clamp and submerged into highly agitated, heated vats of the liquid coating material.

Mechanism: The lab controls the viscosity of the liquid and the precise speed at which the lens is "withdrawn." As the lens rises, gravity ensures the excess fluid drains away, leaving an absolutely uniform layer that is measured in nanometers (Walker, 2014). The lens is then transferred to an oven for several hours (thermal curing), where the heat triggers a chemical reaction that "cross-links" the molecules into their final, hard state.

Spin Coating (UV Cure)

This method is often faster and ideal for single-vision lenses in high-volume settings.

Process: The lens is placed onto a high-speed chuck. A precise drop of the liquid SR material is deposited onto the center of the static lens.

Mechanism: The lens then spins rapidly. Force drives the fluid outward toward the lens edges, spreading it into a micro-thin, perfect layer (Hewitt, 2014). The thickness is controlled by the spin speed and duration. Instead of an oven, this coating is cured by milliseconds of exposure to high-intensity Ultraviolet (UV) light, which instantly "snaps" the molecular chains into a rigid structure.

Chapter Thirteen

The Prescription

The Rx

The Rx from the eye doctor can be presented in many formats but all contain the information needed to make to make a pair of eyeglasses.

		Sphere	Cylinder	Axis	Prism	Base
Distance	OD					
Distance	OS					
Add	OD					
Add	OS					

- OD – Latin for right eye (Oculus Dexter)
- OS – Latin for left eye (Oculus Sinister)
- Sphere – The lens power, in diopters
- Axis – The meridian of the sphere power, if blank the sphere power is present in all meridians
- Cylinder – The power in the meridian 90 degrees from the sphere power
- Prism – the amount of prescribed prism
- Base – the direction of the prism
- Add – The power added to the sphere for the full reading power

Occasionally you will see OU notated. OU is Latin for "both eyes", Oculus Uterque.

The Rx, Plus& Minus Cylinder Forms

In the optical profession two methods of writing the distance prescription are conventional. MDs tend to use the form called plus cylinder. Optometrists tend to use the form lenses are manufactured in, minus cylinder.

		Sphere	Cylinder	Axis	Prism	Base
Distance	OD	-5.00	-1.00	45		
Distance	OS	-5.00	-1.00	135		
Add	OD					
Add	OS					

		Sphere	Cylinder	Axis	Prism	Base
Distance	OD	-6.00	+1.00	135		
Distance	OS	-6.00	+1.00	45		
Add	OD					
Add	OS					

A toric lens has cylinder (the football shape) and modern lenses fabricate that cylinder on the back surface. Many decades ago, it was ground on the front surface. Science learned that the optics were better (closer to the eye) when placed on the back.

Thus, opticians use the minus cylinder form and transpose the occasional plus cylinder Rx into minus cylinder form.

What is the difference in optical power between a plus or minus cylinder Rx? The answer is: **no difference**. It is two ways of writing the same thing!

To convert a "**Plus Cylinder Rx**" to a "**Minus Cylinder Rx**" is very simple with very simple math. There are three steps. Let us convert the Rx **-6.00+1.00 x 135.**

STEP 1 - Add the sphere and cylinder (-6.00 + (+1.00) = -5.00

STEP 2 – Change the cylinder sign from plus to minus.

STEP 3 – Change the Axis Power by 90°

RESULT – **(-5.00-1.00 x 45**)

The Three-Step Transposition Process

Combine the Sphere and Cylinder: (Algebraically) Add the sphere power and the cylinder power to create the new **Sphere**.

Reverse the Cylinder Sign: Change the plus sign (+) to a minus sign (-) while keeping the numerical value. This is the new **Cylinder**.

Rotate the Axis: If the original axis is 90° or less, **add** 90°. If it is greater than 90°, **subtract** 90. This is the new **Axis**.

Multifocal Powers Distance to Near

The Multifocal Rx

		Sphere	Cylinder	Axis	Prism	Base
Distance	OD	+1.25	-0.50	80		
Distance	OS	+1.50	-1.00	75		
Add	OD	+2.00				
Add	OS	+2.00				

The powers, plotted on the optical cross for both the distance Rx and, on the bottom, the near Rx.

Just like the toric SV lens the multifocal maintains the same toricity on the lower section, with the near powers.

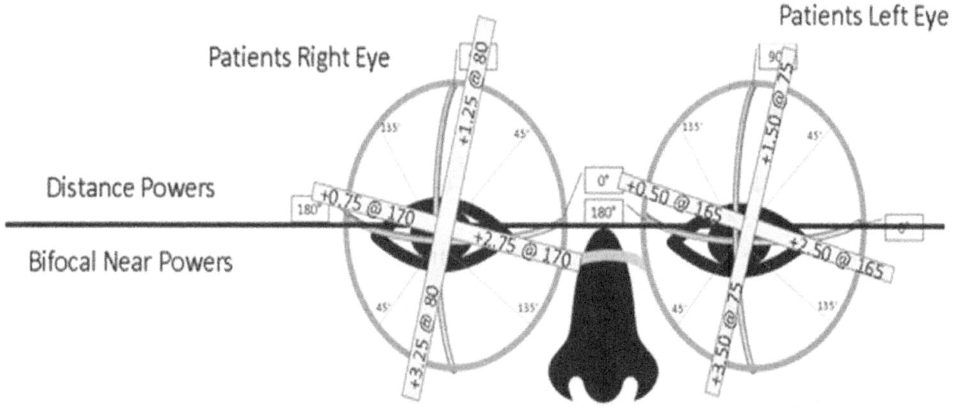

The presence of toric power in a bifocal lens describes the correction for astigmatism, which is typically generated by a "toric surface" featuring two distinct radii of curvature at right angles to each other (Hewitt, 2014). In the vast majority of modern ophthalmic spectacle lenses, the toric correction (the cylinder) is generated onto the back surface of the lens to minimize distortion and maximize the field of view. The bifocal segment, conversely, is typically a "plus" addition generated onto the front surface of the lens. Because the cylinder is applied to the rear surface while the reading addition is applied to the front, the actual **cylinder value remains constant across both the distance portion and the near segment** (Walker, 2014). The toric power does not vary; only the resultant total power shifts between distance, intermediate and near.

Convert Multifocal Rx to Near Only

The Conversion Steps

Take the distance Sphere value and add it to the Add power. This combined value becomes your new Reading Sphere. **+1.25** Sphere combined with**+ 2.00** Add = **+3.25 is new sphere**.

Keep the Cylinder and Axis: **Do not change the Cylinder** (the amount of astigmatism) **or the Axis** (its orientation). These values must remain exactly as they are in the distance prescription to ensure clarity (Serway & Jewett, 2018).

Resulting Rx: The final prescription consists of the new combined Sphere, the original Cylinder, and the original Axis.

The multifocal Rx can be combined into a reading only Rx. Warn the patient that this will only work up close. Distance vision will be blurry with this combination.

Multifocal Rx

		Sphere	Cylinder	Axis	Prism	Base
Distance	OD	+1.25	-0.50	80		
Distance	OS	+1.50	-1.00	75		
Add	OD	+2.00				
Add	OS	+2.00				

Add the Sphere power to the Add power for reading glasses. Maintain the same cylinder and axis.

Converted to Near Rx

		Sphere	Cylinder	Axis	Prism	Base
Near	OD	+3.25	-0.50	80		
Near	OS	+3.50	-1.00	75		
Add	OD					
Add	OS					

Convert Multifocal Rx to Intermediate Rx

To fabricate an intermediate (arm's length) power Rx use exactly the same process but use one-half of the add power when combining with the sphere power.

Rx with Prism

The prism Rx is typically used to correct "double vision" or convergence eyestrain by assisting both eyes to merge the double image into one image.

		Sphere	Cylinder	Axis	Prism	Base
Distance	OD	+1.25	-0.50	80	2	Up
Distance	OS	+1.50	-1.00	75	2	Down
Add	OD					
Add	OS					

		Sphere	Cylinder	Axis	Prism	Base
Distance	OD	+1.25	-0.50	80	2	Out
Distance	OS	+1.50	-1.00	75	2	Out
Add	OD					
Add	OS					

With rare exceptions (see yoked prism) vertical prism should be opposite directions. Horizontal prism should be in the same written direction. (Up & Down, In & In or Out & Out)

Yoked Prism

In ophthalmic dispensing and vision therapy, yoked prism refers to the application of prisms of equal power and identical base direction to both eyes. Unlike conventional prism correction, which is typically used to align eyes that are pointing in different directions (strabismus), yoked prisms do not change the relative position of the eyes to each other. Instead, they shift the entire perceived visual field in a single direction (Goldstein & Cacciamani, 2022).

Prism Direction Rules

The basic rules for splitting prism in an ophthalmic prescription are governed by the principle of **binocular symmetry**. Splitting a large prism value between the two lenses reduces the thickness and weight of a single lens that containss all the prism while minimizing the "prentice effect"; distortion caused by looking through the thicker, wedge-shaped lens (Hewitt, 2014; Walker, 2014).

Rule 1: Splitting Horizontal Prism (In and Out)

Horizontal prism (used for eyes that turn in or out) is always split by maintaining the **same base direction** for both eyes. If a patient requires 6.00 Base Out (BO) total, you split it into 3.00 BO for the right eye and 3.00 BO for the left eye (Walker, 2014). This is the equivalent of 6.00 BO in just one lens.

Logic: Horizontal prism bases are "mirrored" across the bridge of the nose. BO in both eyes pulls the images toward the ears, effectively helping the eyes converge or diverge as a pair (Serway & Jewett, 2018). Geometry defines the alignment.

Rule 2: Splitting Vertical Prism (Up and Down)

Vertical prism (used for one eye that sits or turns higher or lower than the other) is always split by using **opposite base directions** for each eye. If a patient requires 4.00 Base Up (BU) over the right eye, the split equivalent is 2.00 BU in the right eye and 2.00 Base Down (BD) in the left eye (Hewitt, 2014).

Logic: Vertical prism bases are antagonistic. To move the image "up" in one eye, you must move the image "down" in the other to maintain a single, fused binocular image (Serway & Jewett, 2018).

Rule 3: Maintain the Total Power

The "Golden Rule" of splitting is that the **total algebraic sum** of the prism must remain identical to the original prescription. You cannot change the total between the two images; you only change how much of that is corrected by each individual lens (Walker, 2014).

Prism Direction Illustrations

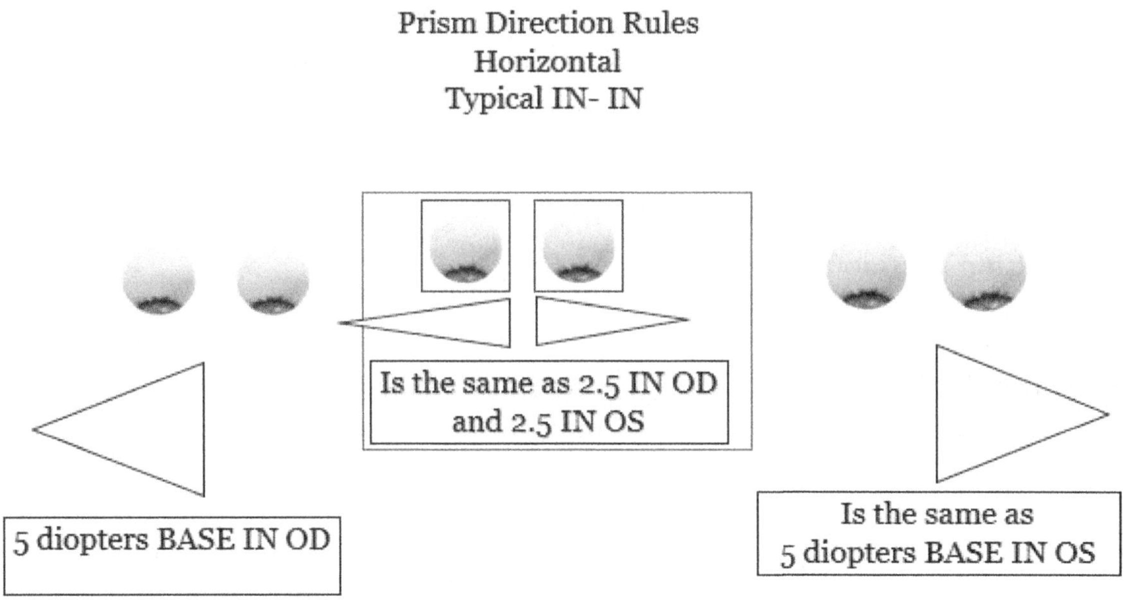

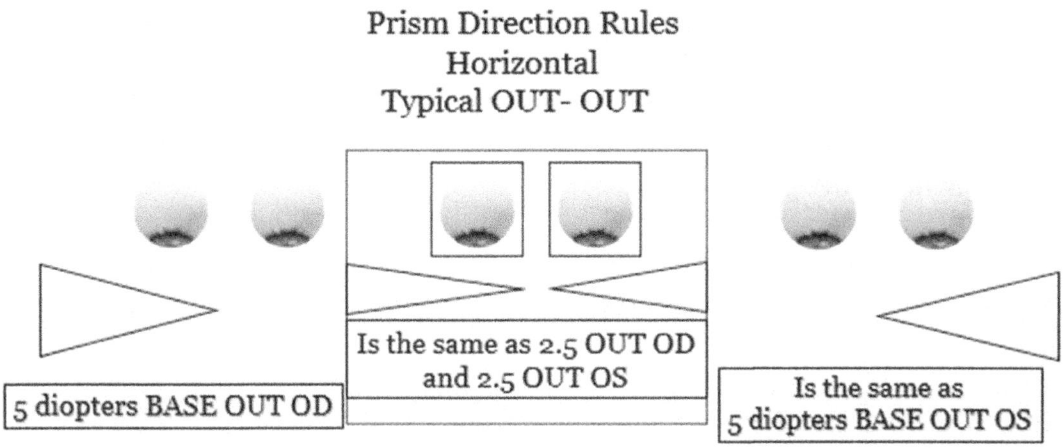

Examples of Horizontal prism; Bases must **be the same** absent yoked prism (very rare and prescribed by doctor).

Prism Direction Illustrations, cont.

Prism Direction Rules
Vertical
Typical UP - DOWN

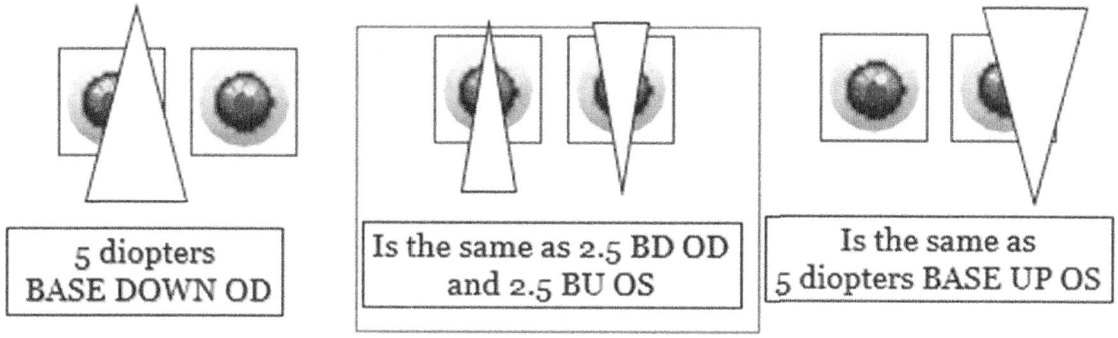

Prism Direction Rules
Splitting

Prism may typically be split for cosmetic reasons. Placing all the prism in one lens may make a very thick heavy lens on one side. Split the prism according to the direction rules to balance the prism.

Turn this:

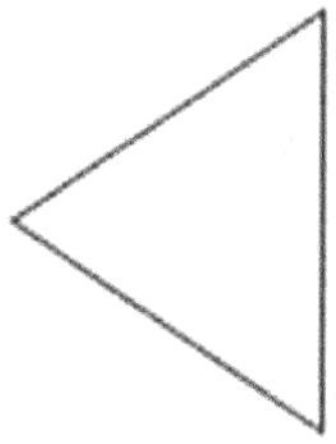

Into this:

Prism Direction Illustrations, cont.

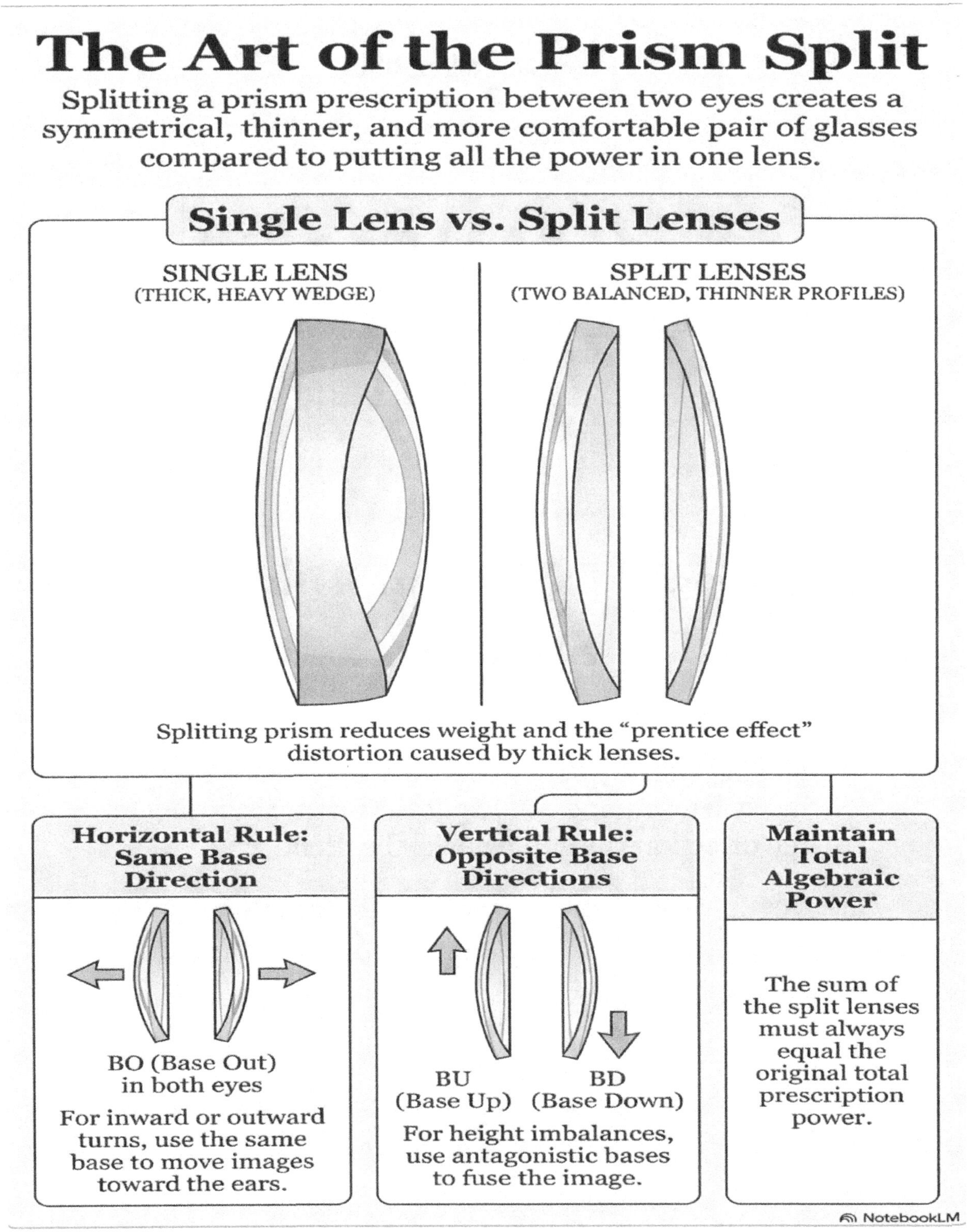

Chapter Fourteen

The PD

IN the world of eyeglasses, **PD** stands for **Pupillary Distance**. This is the specific measurement of the distance between the centers of your two pupils. Because your eyes are dynamic moving parts, this distance actually changes depending on what you are looking at (Hewitt, 2014; Walker, 2014). Movement defines the measurement.

The Distance PD: Looking at the Horizon

When you look at something far away, like a mountain or a movie screen, your eyes point straight ahead. In this position, your pupils are at their widest possible gap. This is called your **Distance PD (DPD)**.

Opticians use this measurement to align the "sweet spot" of the lenses, **the optical center,** directly in front of the patients pupils for activities like driving or walking. If the lenses are off center, it can cause **"unwanted prism,"** making you feel like the world is shifting or giving you a headache (Serway & Jewett, 2018).

The Near PD

When you shift your focus to read a book or look at your phone ,they don't just stay straight; they **converge** (tilt inward) toward each other. This creates a "V-shape" with your line of sight.

Because your eyes have tilted inward, the distance between your two pupils actually **shrinks** (Walker, 2014). This smaller measurement is the **Near PD (NPD)**.

For most people, the Near PD is about **3 mm to 4 mm smaller** than their Distance PD. If you are making reading glasses, the lenses OCs must be closer together to address the convergence (Serway & Jewett, 2018).

Why the Difference Matters

Imagine a pair of binoculars. If you don't adjust the width to match your eyes, you see two weird circles instead of one clear picture. Eyeglasses work the same way.

Distance Glasses: Use DPD so you can see the distance image clearly.

Reading Glasses: Use the NPD so the patients eyes are looking through the OCs of the lens when the eyes converge.

Progressive Addition Lenses: These are high-tech lenses that have a "corridor" that slants inward toward the bottom to match your eyes as they move from distance to reading (Walker, 2014).

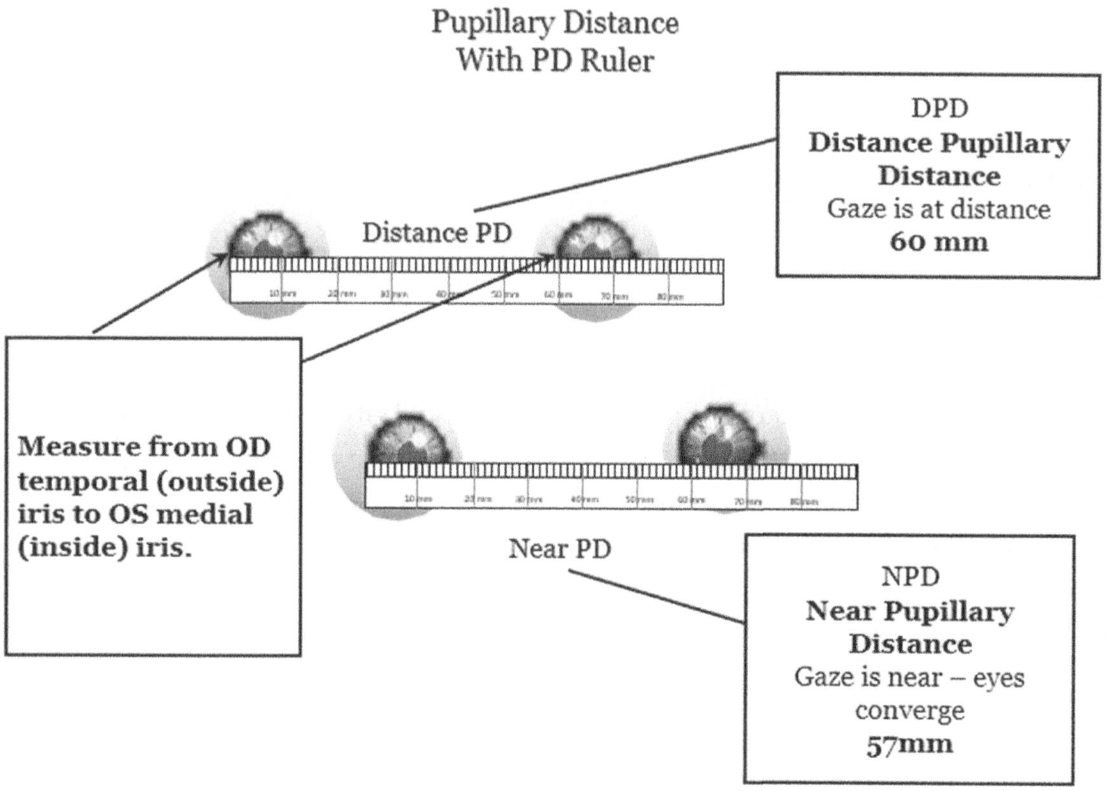

Pupilometer

The pupillometer is the primary tool used in modern ophthalmic dispensing to achieve millimeter precision for both distance and near measurements. It is a high-accuracy digital instrument that eliminates the parallax error (viewing angle error) common with a standard ruler (Hewitt, 2014). This device uses a series of internal mirrors and a light source to simulate different viewing distances (Walker, 2014).

Simulating the Distance PD

When the optician sets the pupillometer to the "infinity" setting, the internal light source is positioned at a focal point that forces the eyes to remain parallel. As the patient looks at the small glowing target inside the device, the pupils are at their widest separation. The optician aligns two vertical hairlines (measuring marks) with the center of the pupils (Serway & Jewett, 2018). This provides the **DPD**.

Simulating the Near PD

To measure your reading position, the optician simply slides a lever on the device to a specific reading distance, typically **35 cm or 40 cm**. Inside the pupillometer, the target physically moves closer to the patient's eyes. This forces their eyes to naturally **converge** (tilt inward) just like they would when reading a book (Walker, 2014). The optician then re-aligns the hairlines to capture the new, narrower distance. This is N**PD**.

Monocular vs. Binocular Measurements

A key advantage of the pupillometer is its ability to provide **monocular** measurements, the specific distance from the center of your nose to each individual eye. Most humans are **not perfectly symmetrical**. One eye might be 32 mm from the nose, while the other is 34 mm (Hewitt, 2014). Using a single "total" PD would result in both lenses being slightly off-center. The pupillometer ensures each lens is custom-aligned to its respective eye. Accuracy facilitates the focus.

Pupillary Distance
With Pupilometer

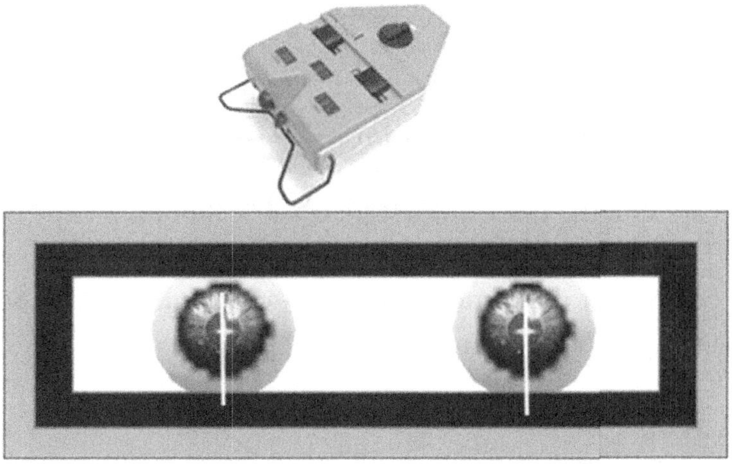

With the pupilometer you may measure either the DPD or the NPD. Instruct the patient to fixate on the internal light. Slide the bars over the reflection on each eye and read the external gauge. This instrument will measure MONOCULAR PD. Using the PD for *each eye separately* is more accurate than using the BINOCLAR PD taken with the PD Stick.

State-of-the-Art Measurements

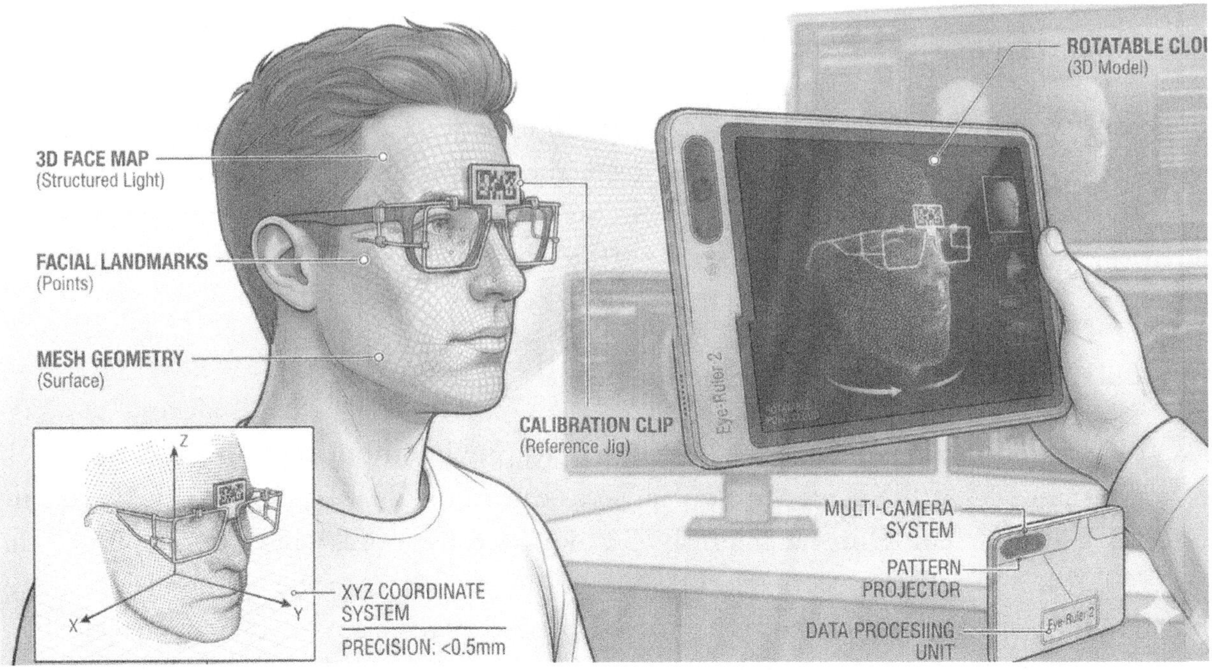

The contemporary ophthalmic dispensing environment has transitioned from manual rulers to sophisticated **Digital Imaging Systems** (DIS), typically hosted on high-resolution tablets like the iPad. These systems utilize the tablet's camera and specialized reference attachments to capture a comprehensive "optical map" of how a specific frame sits on a specific face (Hewitt, 2014; Walker, 2014).

The Reference Frame: The Key to Precision

To use a tablet-based system, the optician first places a "calibration clip" or reference jig onto the patient's chosen frame. This clip contains known physical dimensions that the software uses as a baseline to correct for the camera's distance and angle (Serway & Jewett, 2018). The tablet then captures a single front-view and side-view photograph.

The Five Essential Measurements

Unlike a manual pupillometer, which only measures the horizontal gap, a digital system captures a 3D data set required for high-definition free-form lenses:

Monocular Pupillary Distance (PD): The horizontal distance from the bridge of the nose to the center of each pupil.

Fitting Height (Ocular Center): The vertical distance from the bottom of the lens to the center of the pupil. This is critical for progressive lenses to ensure the "corridor" starts at the right height (Walker, 2014).

Vertex Distance: The precise distance from the back surface of the lens to the front of the cornea. High prescriptions require power adjustments if this distance changes (Serway & Jewett, 2018).

Pantoscopic Tilt: The angle at which the frame leans toward the cheeks. Digital surfacing uses this to optimize the lens for the wearer's specific "looking-down" angle.

Face Form (Wrap): The horizontal curvature of the frame across the face. This measurement is used to eliminate peripheral distortion in wrap-around sunglasses (Hewitt, 2014).

The Science of Optimization

Modern free-form digital lenses are manufactured using a process called **Position of Wear (POW)** optimization. When the tablet captures these five measurements, the data is sent to the lens laboratory's computer. The lab's algorithm then re-calculates the prescription to account for exactly how the lens is tilted and spaced on the wearer's face (Walker, 2014). Without these digital measurements, a high-tech lens is essentially performing at only a fraction of its potential.

Chapter Fifteen

Prentice's Rule and Axis Power Determination

Prentice's Rule

AS we have learned, the distance between the centers of your patient's eyes is called the pupillary distance. This measurement must match the distance between the optical centers of the lenses you make. Errors will cause eyestrain, blurry vision, headaches and unhappy patients (followed by unhappy opticians). When you inspect the final product, you will employ the concept of **Prentice's Rule** to determine pass or fail.

Prentice's Rule is a simple math formula that opticians use to figure out how much "prism" is created when your eye isn't looking through the exact center of a lens. Every eyeglass lens is like two prisms joined together, and the very middle is the "sweet spot" where the light doesn't bend at all.

Prentice's Rule

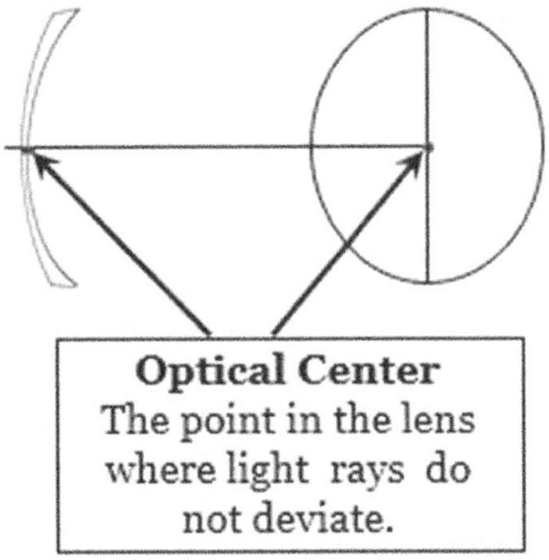

P = prism in diopters
C = decentration in mm
D = lens power in diopters

$P = (C)(D)/10$

Use Prentice's Rule to determine how much prism is created when the OC is placed at a point that does not line up with the patients PD. In some cases this can be used to decenter a lens to create the prescribed prism without special grinding.

Prism power equals the decentration in millimeters multiplied by the diopter power of the lens. There are occasions when prescribed prism can be filled simply by **decentering** the lens during fabrication as opposed to having it ground" into the lens. This can save time and dollars. Prentice's Rule is an optician superpower.

Decentration with Prentice's Rule

Prentice's Rule

	Sphere	Cylinder	Axis	Prism	Add
OD	-5.00				
OS	-5.00				

Patient PD = 60 mm

Distance between OCs = 68 mm

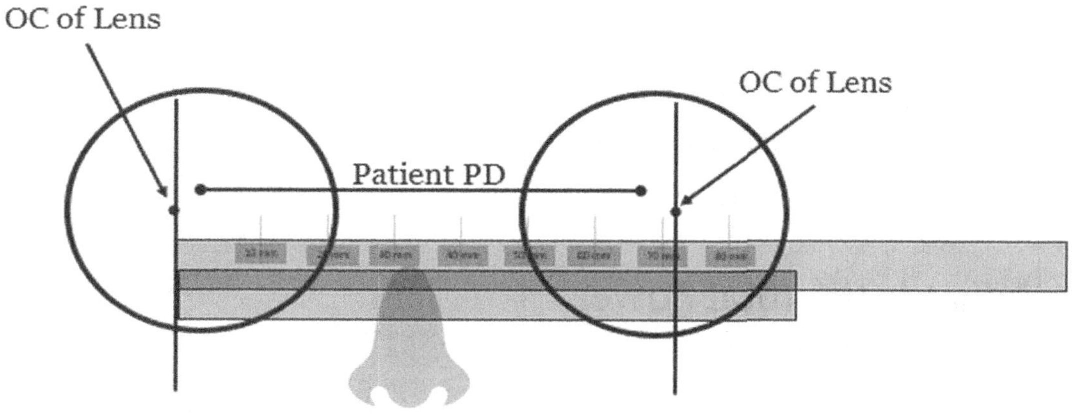

Question:
If the patient PD is 60 mm and the distance between the OC's is 68 mm; How much unwanted prism is created in the -500 OU Rx?

P = (C)(F)/10

P = prism in diopters
d= decentration in mm
D = lens power in diopters

C= 8 mm
D = 5.00
P = 8 x 5/10

P = 8 x .5

P = 4D

The same process may be used for each single lens with a monocular PD.

Axis Power Determination

Axis Power Determination
With-the -Rule

	Sphere	Cylinder	Axis	Prism	Add
OD	-5.00	-2.00	180		
OS	-5.00	-2.00	180		

Question:
What is the total power at axis 180 of each eye?

What is the total power at axis 90 of each eye?

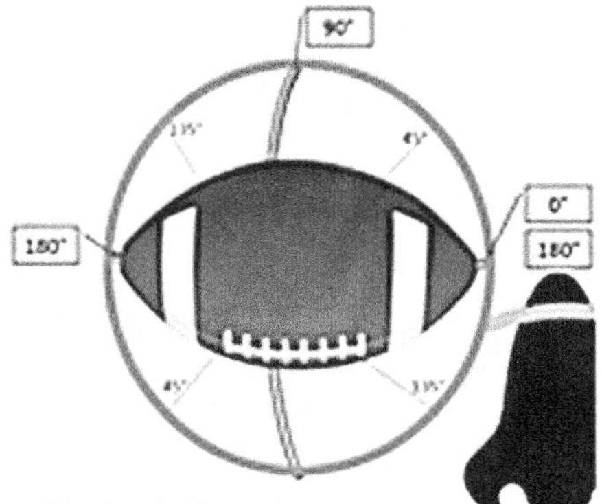

With-the-Rule Astigmatism

Answer:
The total power at 180 is the sphere power (-5.00).

The total power at 90 is the sphere power PLUS the cylinder power (-7.00).

Axis Power Determination cont.

Axis Power Determination
With-the -Rule

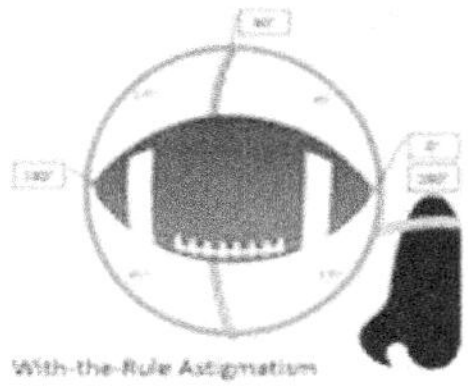

	Sphere	Cylinder	Axis	Prism	Add
OD	-5.00	-2.00	180		
OS	-5.00	-2.00	180		

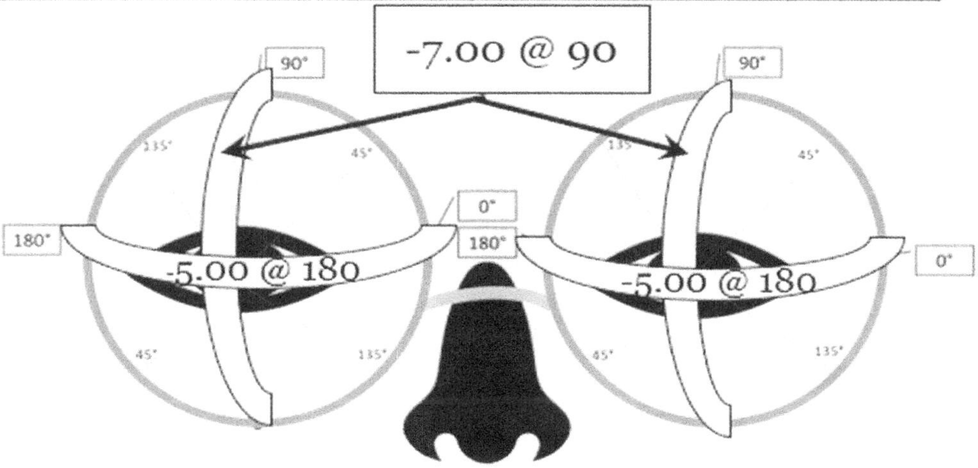

Question:
What is the total power at axis 180 of each eye?
What is the total power at axis 90 of each eye?

Note the location of the powers on the optical cross compared to the Rx.

@180 > -5.00

@90 > -7.00

Axis Power Determination cont.

Axis Power Determination
Against-the-Rule

	Sphere	Cylinder	Axis	Prism	Add
OD	-5.00	-2.00	**90**		
OS	-5.00	-2.00	**90**		

Question:

What is the total power at axis 180 of each eye?

What is the total power at axis 90 of each eye?

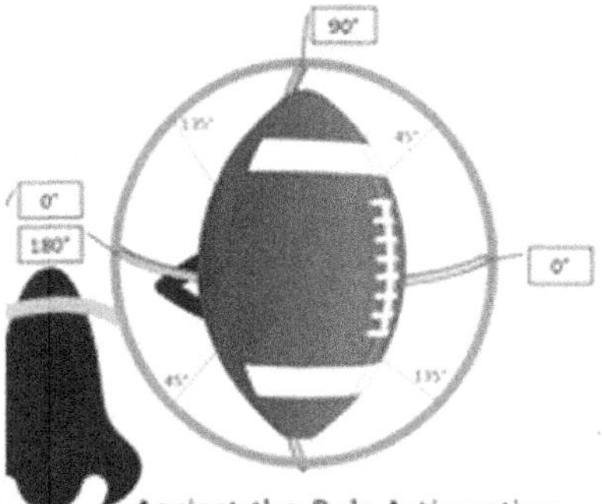

Against-the-Rule Astigmatism

Answer:

The total power at 180 is the **sphere** power **PLUS** the **cylinder** power (-7.00).

The total power at 90 is the **sphere** power (-5.00).

Axis Power Determination cont.

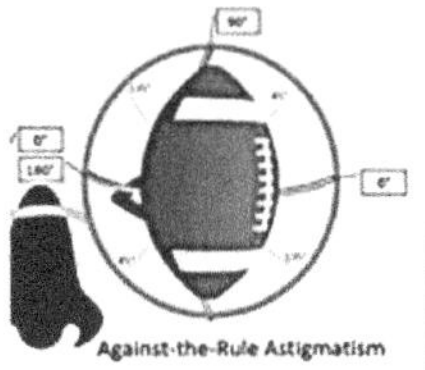

Axis Power Determination
Against-the-Rule

	Sphere	Cylinder	Axis	Prism	Add
OD	-5.00	-2.00	90		
OS	-5.00	-2.00	90		

Question:

What is the total power at axis 90 of each eye?

What is the total power at axis 180 of each eye?

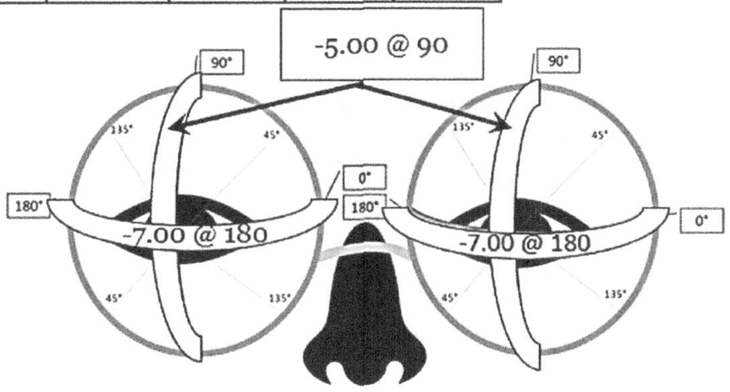

Axis Power Determination

Key Takeaway for Axis Determination

1. The sphere power is at the prescribed axis.
2. The cylinder power PLUS the sphere power is 90 degrees away on the optical cross.
3. Add the sphere and the cylinder algebraically.

For example:

3.00 + 3.00 = 6.00

3.00 + (-3.00) = 0.00

Axis Power Determination Oblique

Axis Power Determination
Oblique

	Sphere	Cylinder	Axis	Prism	Add
OD	-5.00	-2.00	135		
OS	-5.00	-2.00	135		

Question:
What is the total power at axis 135 of each eye?

What is the total power at axis 45 of each eye?

Answer:
The total power at 135 is the sphere power (-5.00).

The total power at 45 is the sphere power PLUS the cylinder power (-7.00).

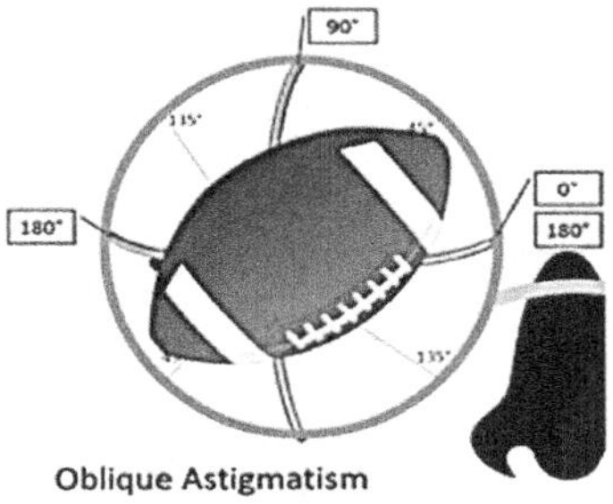

Oblique Astigmatism

Axis Power Determination
Oblique

	Sphere	Cylinder	Axis	Prism	Add
OD	-5.00	-2.00	135		
OS	-5.00	-2.00	135		

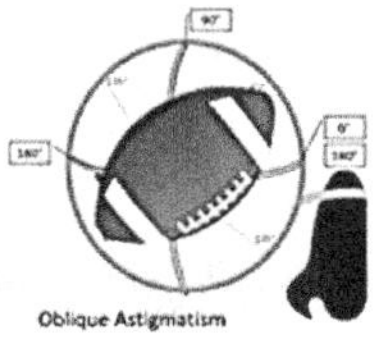

Question:

What is the total power at axis 135 of each eye?

What is the total power at axis 45 of each eye?

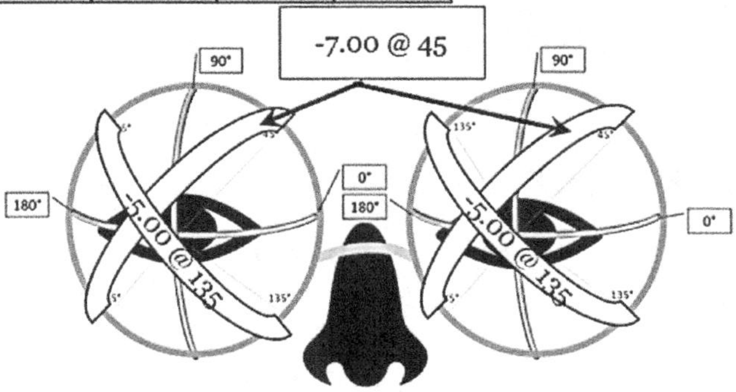

Axis Power on 180 line with

Axis 180 Power Calculation
With Oblique Axis

	Sphere	Cylinder	Axis	Prism	Add
OD	-5.00	-2.00	135		
OS	-5.00	-2.00	135		

Question:
Why is it important to determine total power at axis 180 of each eye?

Answer:
To employ Prentice's Rule during fabrication and final inspection of lenses. Wanted or unwanted prism can be determined.

Axis 180 Power Calculation
With Oblique Axis

	Sphere	Cylinder	Axis	Prism	Add
OD	-5.00	-2.00	135		
OS	-5.00	-2.00	135		

Question:

What is the total power at axis 180 of each eye?

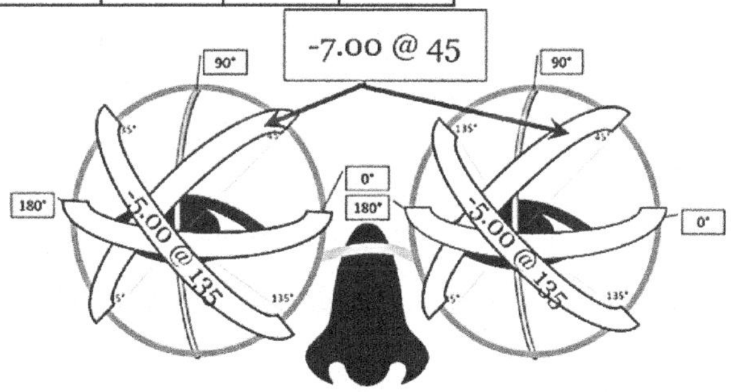

Power on 180 = Cylinder Power (lessor of degrees to 180 or zero/90) + sphere power

Step One: Take the number of degrees to the 180 (or zero) line and divide by 90.
Step Two: Take the Cylinder power and multiple by the percentage from Step One.
Step Three: Add the Step Three result to the sphere power.

Result is the power on the 180 line. Step 1: 45°/90°= .5 Step 2: -2.00 * .5 = 1.00
Step 3: -5.00 + (-1.00) = -4.00

The sphere power on the 180 line is -4.00

Axis 180 Power Cheat Sheet

Note how 45 degrees away from the 180 is one-half the cylinder power added to the sphere power. The percentage of the cylinder needed to get to the 180 or 0-degree line, added to the sphere gives you the power on the 180.

Cylinder Degrees from 180	Formula	Percentage
90	90/90	1
85	85/90	.94
80	80/90	.89
75	75/90	.83
70	70/90	.78
65	65/90	.72
60	60/90	.67
55	55/90	.61
50	50/90	.56
45	45/90	.50
40	40/90	.44
35	35/90	.39
30	30/90	.33
25	25/90	.25
20	20/90	.22
15	15/90	.17
10	10/90	.11
5	5/90	.056

Prentice's Rule with 180 Line Calculated

Prentice's Rule
With 180 Line Calculated

An optician must be able to calculate the power on the 180 line and then determine, using Prentice's Rule, **the amount of unwanted or wanted prism in a lens.**

JOB #1

180 Line Calculation

$C(d/90) + S = P$

P = Power on 180
C = Cylinder Power
d = Cyl degrees to 0 or 180, from the written cylinder and always equal to or less than 90
S = Sphere Power

+

JOB #2

Prentice's Rule

$(C)(D)/10 = P$

P = prism in diopters
C = decentration in **mm**
D = lens power in diopters

Caution: On occasion you will see **Prentice's Rule applied in centimeters** as opposed to millimeters. Use the formula the same way **without dividing by ten** as illustrated.

THE PAINED PATIENT

Once upon a time, a weary man named Arthur walked into the neighborhood optical shop holding his head in pain. Every single time he wore his brand new glasses, from an online bargain shop, terrible headaches started. He felt miserable. The kind optician listened to Arthur explain his frustrating problem with the new glasses

The optician took the glasses and placed them into the lensometer that read exactly where the OCs were located. She discovered that the OCs of the lenses were placed much too far apart.

Because the light was bending incorrectly, Arthur's eye muscles were forced to work incredibly hard just to see things. This constant heavy straining was the reason behind his terrible daily headaches. The optician knew exactly how to fix the uncomfortable glasses for the tired man.

She quickly ordered a brand new pair of lenses (from an excellent lab that offered no charge remakes for their accounts, and yes, that lab actually exists) with the centers measured perfectly to match the space between his eyes. When the new glasses arrived, Arthur put them on his face with hope. The headaches were gone. He could read his favorite books again with clear and comfortable vision. That is service with a Cherry (lab) on top. THE END

Prentice's Rule with 180 Line Calculated, cont.

Prentice's Rule
With 180 Line Calculated

Job #1 – Calculate the powers on the 180 line for both lenses.

Question:
If the patient PD is 60 mm and the distance between the OCs of the lens is 68 mm; How much unwanted prism is created?

	Sphere	Cylinder	Axis	Prism	Add
OD	-5.50	-1.75	95		
OS	-5.00	-2.25	35		

P = C (d/90) + S

P = Power on 180
C = Cylinder Power
d = Cyl degrees to 0 or 180, less than 90
S = Sphere Power

OD:
- P = -1.75 (85/90) + (-5.50)
- P = -1.75 (.94) + (-5.50)
- P = -7.15 @ 180

85 degrees to the 180 line

OS:
- P = -2.25 (35/90) + (-5.00)
- P = -2.25 (.39) + (-5.00)
- P = -5.88 @ 180

35 degrees to the 180 line

Prentice's Rule
With 180 Line Calculated

Job #2 – Calculate the unwanted prism for each lens and the total unwanted prism for both lenses together. D will equal the power in the 180.

Question:
If the patient PD is 60 (30/30) and the distance between the OCs of the lens is 68 (34/34) mm; How much unwanted prism is created?

	Sphere	Cylinder	Axis	Prism	Add
OD	-5.50	-1.75	95		
OS	-5.00	-2.25	35		

Prentice's Rule

P = prism in diopters
C = decentration in mm
D = lens power in diopters

P = (C)(D)/10

OD:
- P = -1.75 (85/90) + (-5.50)
- P = -1.75 (.94) + (-5.50)
- P = -7.15 @ 180

OS:
- P = -2.25 (35/90) + (-5.00)
- P = -2.25 (.39) + (-5.00)
- P = -5.88 @ 180

OD:
- P = (4)(-7.15) /10
- P = 2.86

OD:
- P = (4)(-5.88) /10
- P = 2.35

Unwanted Prism

Prism, Unwanted vs Wanted

Prentice's Rule
With 180 Line Calculated

Question:
What is the direction of the unwanted prism?

	Sphere	Cylinder	Axis	Prism	Add
OD	-5.50	-1.75	95		
OS	-5.00	-2.25	35		

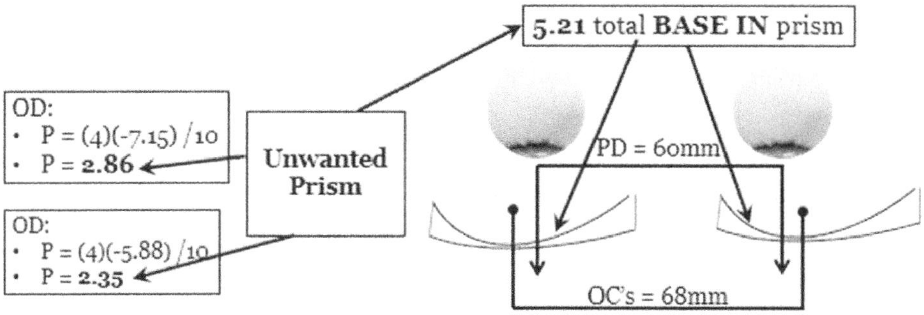

Prentice's Rule
With 180 Line Calculated

Question:
If the doctor prescribed BASE IN prism with this Rx, could you obtain that prism by decentering the lens to obtain WANTED prism?

	Sphere	Cylinder	Axis	Prism	Add
OD	-5.50	-1.75	95	2.5 IN	
OS	-5.00	-2.25	35	2.5 IN	

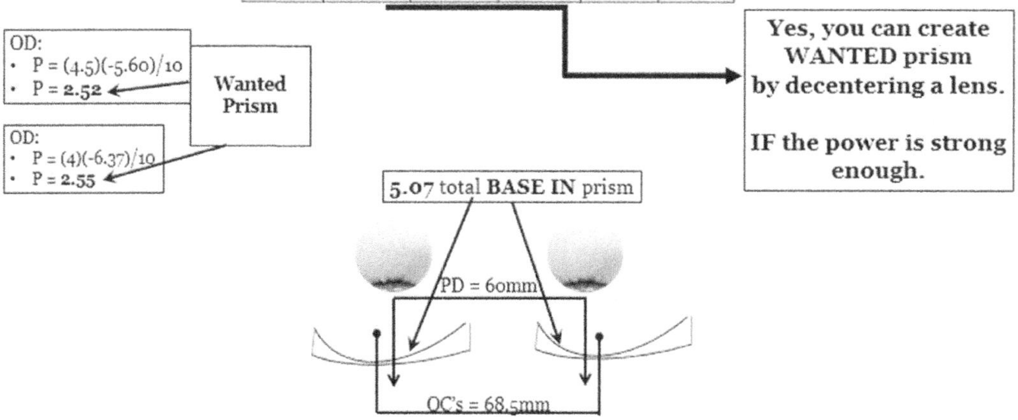

The Value of Prentice's Rule

Determining the power along the **180 line** is a foundational step in ophthalmic dispensing that allows for the calculation of horizontal prismatic effects. By applying **Prentice's Rule**, an optician can identify whether a lens setup creates **wanted prism** (prescribed to correct eye alignment) or **unwanted prism** (induced by errors that cause visual strain) (Hewitt, 2014; Walker, 2014). Accuracy governs the binocular experience.

The 180 Line and Horizontal Power

In a standard eyeglass prescription, the lens power often varies across different orientations (meridians) due to astigmatism correction. The 180 line represents the horizontal plane passing through the optical center of the lens.

Sphere Lenses: The power is identical in all directions, so the 180 line power is simply the sphere value.

Cylinder Lenses: The power at 180 line must be calculated based on the axis. If the axis is at 90°, the full cylinder power is active along the 180 line. If the axis is at 180, only the sphere power exists on that line (Serway & Jewett, 2018).

Determining this specific horizontal power is the "input" required to solve Prentice's Rule for side-to-side eye movements.

Applying Prentice's Rule

Prentice's Rule provides the mathematical bridge between lens power (D) and the physical distance (C in mm) a pupil is moved away from the optical center.

Unwanted Prism: This occurs when a lens is manufactured with the wrong **Pupillary Distance (PD)**. If a lens is "decentered" because it doesn't line up with the eye, it induces a prism that the brain must work to overcome. For a patient with high power, even a 2 mm error can create enough unwanted prism to cause headaches or "pulling" sensations (Walker, 2014).

Wanted Prism: Sometimes, a doctor deliberately "decenters" a lens to create a specific prismatic effect without using a heavy, wedge-shaped prism. By moving a standard lens slightly toward the nose or temple, the optician uses the power of the lens to provide the prescribed correction (Serway & Jewett, 2018).

Chapter Sixteen

Eyeglass Frame Measurements

A, B & ED

IN ophthalmic dispensing, the **Boxing System** provides a standardized matrix of measurements to ensure that a lens is precisely centered within a chosen frame. This system relies on three primary dimensions—**A**, **B**, and **ED**—to define the boundaries of the lens aperture and the placement of the optical center relative to the wearer's pupil. These measurements are the geometric foundation of a successful prescription.

The A Dimension and the Datum Line

The **A dimension** represents the maximum horizontal width between the vertical borders of the lens box. It is measured from the **inner grooves** of the frame to account for the full area the lens will occupy.

The Datum Line: This is an imaginary horizontal line that runs exactly halfway between the top and bottom of the lens box.

Importance: The intersection of the Datum Line and the vertical centerline of the "A" box marks the **Geometric Center** of the frame.

Centration: Opticians use the A dimension to calculate the **Frame PD** (A + Bridge size), which is then compared to the patient's actual PD to determine how much horizontal decentration is required to align the lens optics.

The B Dimension

The **B dimension** is the maximum vertical height of the lens box, measured from the highest point to the lowest point of the lens aperture.

Vertical Alignment: While the A dimension handles horizontal placement, the B dimension is critical for vertical positioning.

Multifocal Placement: For bifocals or progressive lenses, the B dimension determines if the frame is deep enough to hold the entire "reading corridor".

Segment Height: Opticians measure the "Seg Height" starting from the bottom of the B dimension up to the pupil (PALs) or lower eyelid (lined segments) Note: Positioning can be variable based on patient needs . If the B dimension is too small, the reading portion of the lens may be cut off too much during manufacturing.

The ED (Effective Diameter)

The **Effective Diameter (ED)** is defined as twice the distance from the geometric center of the lens box to the furthest edge of the lens aperture.

The Diagonal Factor: Because many frames are not perfect rectangles (such as aviators or cat-eyes), the widest part of the frame is often on a diagonal rather than a straight horizontal line.

Lens Blank Size: The ED is the most important number for determining the **minimum lens blank size (MBS)** required.

Avoiding "Cut-Out": If an optician chooses a lens blank that is smaller than the **ED (plus any necessary decentration**), the lens will not physically fill the frame, a failure known as "rolling off" or "cutting out".

Frame Measurements

Frame dimensions are used for the proper placement or glazing of the prescription lenses.

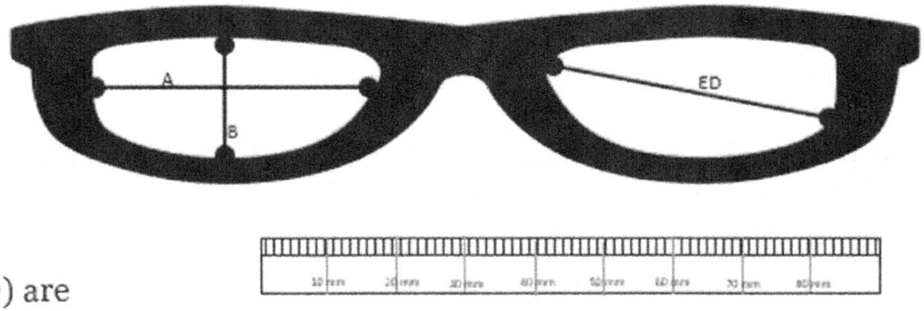

The datum line (or A dimension), the B dimension and the effective diameter (ED) are needed to make lenses properly.

A, B & ED, cont.

When measuring the frame dimensions take into account the groove that holds the lens or add at least one mm to the visible opening.

A = 43 mm

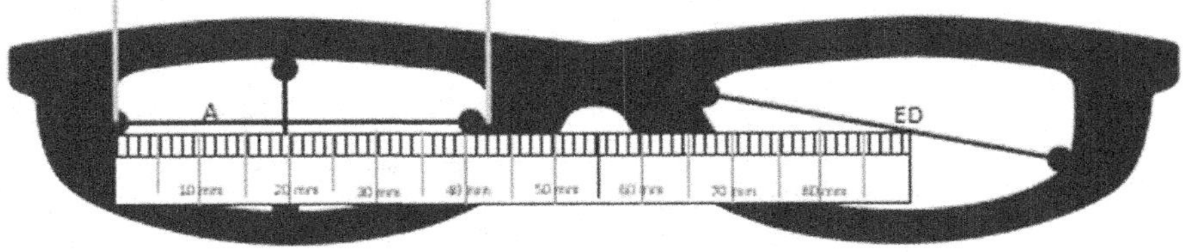

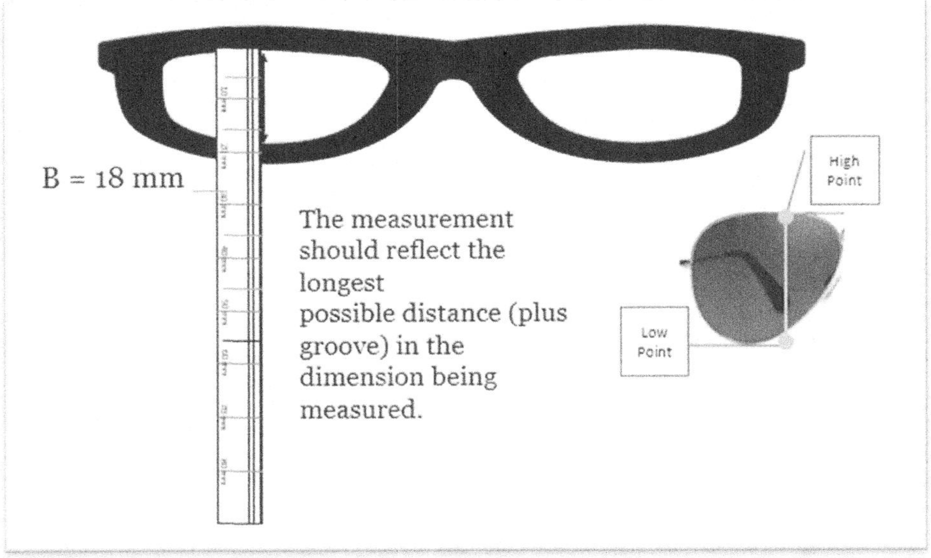

The effective diameter (ED) is the longest possible distance within that frame styles lens opening.

ED = 46 mm

Box System

Boxing System

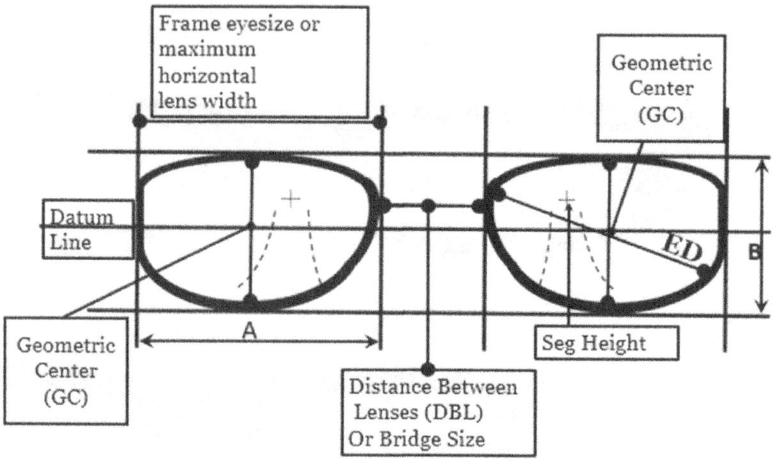

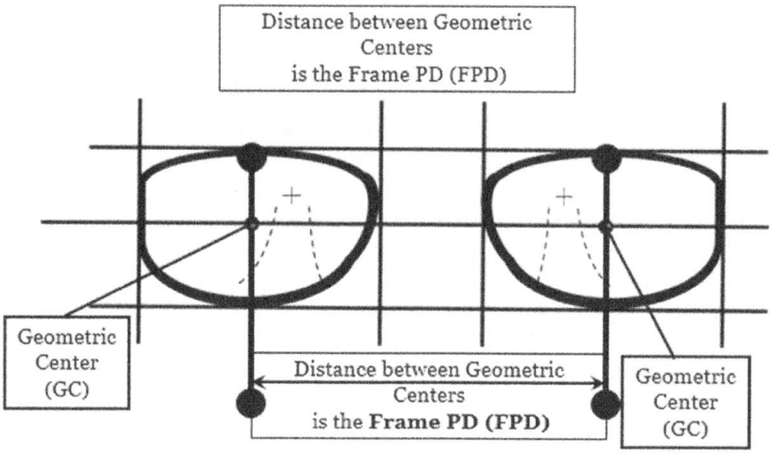

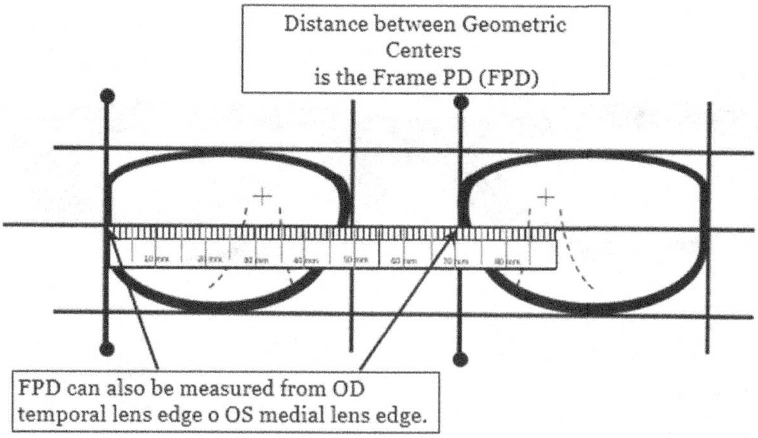

Pulling the PD, OC and FPD together!

Lenses for your patient are either pre-made stock lens blanks or custom made by the lab your company uses. Pre-made single vision lenses are typically available in blank diameters of 65mm, 68mm, 70mm, 72mm, 75mm 80mm and 85mm. Not every type of lens is available every blank size. Also, typically, the smaller the blank size used the thinner and lighter the final product will be and that is a desirable result.

An optical needs to understand how to calculate the minimum blank size required for any combination of frame size and PD for stock SV lenses. The lab software will typically determine multifocal data.

FPD – Frame Pupillary Distance
PPD – Patient Pupillary Distance
ED – Effective Diameter
MBS – Minimum Blank Size

The MBS formula is (FPD – PPD)/2 + ED = MBS)

Minimum Blank Size (MBS)

Step One - Measure the Patient PD (60mm)

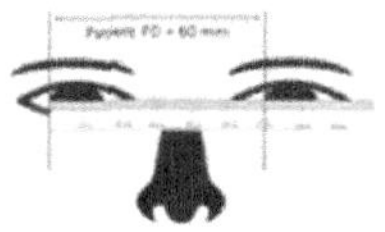

Step Two – Measure the Frame PD (72mm)

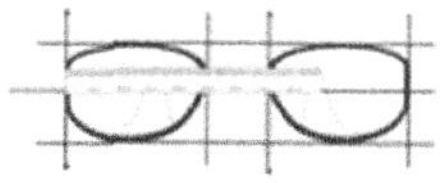

Step Three – Measure the ED (52mm)

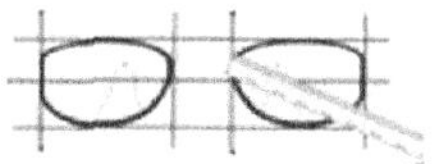

Step Four – Calculate MBS
- FPD – PPD)/2 + ED = MBS
- (72 – 60)/2 + 52 = MBS
- 12/2 + 52 = MBS
- **58mm = MBS**

EXAMPLE

Chapter Seventeen

Lensometer

THE **Lensometer (Vertometer, Focimeter)** is the primary diagnostic instrument used by an ophthalmic dispenser to verify the optical properties of a spectacle lens (Hewitt, 2014). It functions by neutralizing the power of the lens through a series of internal optics, allowing the operator to pinpoint the exact sphere, cylinder, and axis of a prescription (Walker, 2014). This verification process ensures that the finished eyewear matches the doctor's written orders with mathematical certainty.

Verifying the Prescription Power

The core function of the lensometer is to measure the **dioptric power** of a lens. By placing the lens against the instrument's aperture and adjusting the power drum, the dispenser can identify the "sphere" and "cylinder" components of the prescription (Serway & Jewett, 2018).

Sphere Power: The lensometer identifies the point of focus where the primary meridians are neutralized (Walker, 2014).

Cylinder and Axis: For patients with astigmatism, the device measures the difference in power between the two primary meridians and calculates the precise angle (axis) at which the cylinder is oriented (Hewitt, 2014).

Locating the Optical Center (OC)

Every lens has a specific "sweet spot" called the **Optical Center (OC)**, where light passes through without being bent or distorted (Hewitt, 2014). The lensometer allows the dispenser to find this point by centering the internal "target" (the mires) within the reticle's crosshairs.

Marking the Lens: Once the OC is located, the dispenser uses the instrument's built-in ink pins to mark the lens (Serway & Jewett, 2018).

Alignment: These marks are then compared to the patient's **Pupillary Distance (PD)** to ensure the eyes are looking through the correct part of the glass (Walker, 2014). Alignment protects the vision.

Detecting and Measuring Prism

If a doctor has prescribed **prism** to help a patient's eyes work together, or if a lens has been made incorrectly, the lensometer will reveal it.

Prismatic Displacement: When a lens contains prism, the internal target in the lensometer will "jump" or shift away from the center of the reticle (Serway & Jewett, 2018).

Measurement: The rings within the lensometer's view (the reticle) are calibrated in **Prism Diopters**, allowing the dispenser to read exactly how much prism is present and in what direction (Base In, Out, Up, or Down) (Hewitt, 2014).

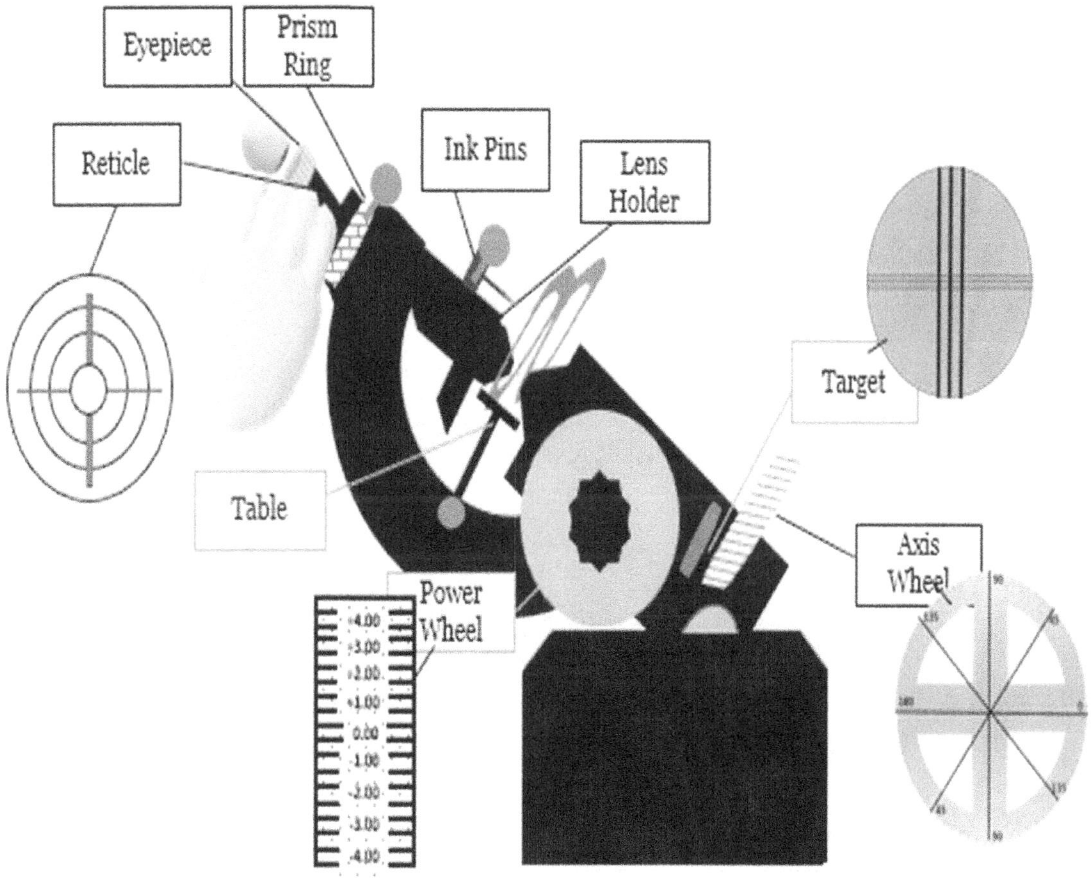

Lensometer, cont.

For an ophthalmic dispenser, the manual lensometer is an indispensable tool, serving as the definitive method for neutralizing, verifying, and marking spectacle lenses. It uses a series of precision internal optics to measure the dioptric power, identify astigmatism correction (cylinder and axis), and detect prismatic effect with mathematical certainty (Hewitt, 2014;

Walker, 2014). Understanding each component and its function is vital for accurate, reliable measurements.

Axis Wheel

The **Axis Wheel** is a rotatable drum located on the side of the lensometer, typically calibrated in degrees from 1 to 180. Its primary function is to rotate the internal target (the luminous mires) to find the precise orientation of the cylinder power for patients with astigmatism. Once the primary sphere power is neutralized, the operator uses the Axis Wheel to align the target lines with the new, secondary meridian's focus (Hewitt, 2014). The final reading on the wheel is the prescription's **Axis**.

Eyepiece (Focusing Requirement)

The **Eyepiece** is the component the operator looks through. It is arguably the most critical and often overlooked part of the entire process. Before taking *any* measurements on a lens, the operator must focus the eyepiece for their individual vision.

Procedure: To do this, look into the instrument without a lens in place. Turn the eyepiece all the way out (counter-clockwise) until the internal reticle (the crosshairs) is blurry. Slowly turn it back in (clockwise) until the reticle just becomes sharp and distinct.

Why it matters: Failing to do this introduces **parallax error** (a fake shift in position due to viewing angle) or forces your own eye muscles to compensate, leading to incorrect power readings, especially in high prescriptions or when measuring prism (Hewitt, 2014). Focusing ensures the operator's eye is a stable baseline.

Ink Pins (Marker Pins)

The **Ink Pins** (or marker pins) are typically a set of three pins that can be depressed to make a physical ink mark on the lens. These marks are essential for subsequent steps in the dispensing process.

Optical Center (OC): The center pin marks the exact point of the lens's Optical Center, where no prism exists (Hewitt, 2014).

Axis Alignment: The other two pins (marking 180 and 0) define the horizontal axis of the lens, which is critical for correctly orienting the cylinder power in the frame (Walker, 2014).

Lens Holder (or Lens Stop)

The **Lens Holder** is a precision aperture that supports the lens during measurement. It has a spring-loaded clamp that must be gently placed on the lens surface to hold it securely. The position where the lens physically touches the holder—the **lens stop**—is the calibrated plane for all power measurements. It is critical to center the lens as much as possible *before* applying the clamp to avoid creating an initial prismatic shift.

Power Wheel

The **Power Wheel** (or power drum) is the main control on the side of the lensometer, typically calibrated in **Diopters** (D) from +20.00D to -20.00D (typically with intermediate markings for every quarter or eighth diopter). This wheel moves the internal target back and forth until its image is brought into clear focus by the lens being measured.

Neutralization: The first point of clear focus identifies the **Sphere Power**. For a lens with astigmatism, a second, weaker point of focus will appear, and the difference between these two readings is the **Cylinder Power** (Hewitt, 2014).

Prism Ring is a component often located below or around the eyepiece. It allows the operator to introduce a known amount of prism into the visual field to neutralize or measure any prism present in the lens. This is especially useful for measuring high amounts of prescribed prism (often up to 25D). If the lens has no prescribed prism and is correctly centered, the internal target will appear in the absolute center of the reticle; any deviation indicates induced prism, which the Prism Ring can quantify (Serway & Jewett, 2018).

The **Reticle** is the fixed internal grid and scale seen through the eyepiece. It serves as the reference against which the luminous target is measured. A typical reticle includes a set of concentric circles calibrated in **Prism Diopters** (D) (e.g., 1, 2, 3, etc.) radiating outwards from the center. It also features horizontal and vertical crosshairs. The lensometer is at its absolute center and in focus when the target is clear and perfectly aligned with the intersection of these crosshairs (Serway & Jewett, 2018).

The **Table** (or lens table) is a flat, horizontal platform designed to support the bottom edge of a spectacle frame or lens during measurement. Many modern tables have a built-in vertical rest, or "frame rest" attachment, which ensures the frame sits evenly on the datum line. The height of this table can be precisely adjusted so that the operator can consistently measure the center of the lens blank.

The **Target** (or mires) is the luminous pattern generated by the lensometer's light source. The clarity, orientation, and focus of this pattern are the key visual feedback for all power and axis measurements.

Procedure: When looking through the eyepiece, the target is adjusted using the **Power Wheel** and **Axis Wheel** until it is in perfect focus.

Sphere Meridian: Focus is achieved when one set of mires (usually parallel lines) is completely sharp. This point on the power wheel defines the **Sphere Power**.

Cylinder Meridian: For astigmatism, a second meridian must be neutralized. A different set of lines (usually parallel dots or a different line orientation) will come into focus when the axis wheel is rotated and the power wheel is adjusted, defining the **Cylinder Power** (Hewitt, 2014; Serway & Jewett, 2018).

Strategic Coordination

The manual lensometer requires the synchronized action of these components to work. The optician looks through the **EYEPIECE** while manipulating the **POWER WHEEL** and **AXIS WHEEL** with one hand, and adjusting the **LENS HOLDER** and **TABLE** with the other to align the **TARGET** precisely with the **RETICLE** crosshairs. Finally, the **INK PINS** mark the definitive optical map of the finished lens blank (Walker, 2014). This coordinated process provides verification of the required optical standard.

Chapter Eighteen

ANSI Standards for the Optician

THE regulatory framework governing optical fabrication relies upon a framework of "tolerances" to prevent physiological harm and ensure visual acuity. Navigating these parameters requires adherence to established standards (American National Standards Institute [ANSI], 2020a).

Standards can change. The Vision Council maintains a quick reference quide that should hang by every opticians lensometer. Use a search engine to locate printable standards guide.

1. ANSI Z80.1: Prescription Ophthalmic Lenses

The primary foundational document for daily dispensary operations dictates the acceptable deviations for sphere, cylinder, and axis power in custom spectacle lenses (ANSI, 2020a). This specific criterion establishes the legal threshold for rejecting noncompliant laboratory work before patient delivery. Precision strictly protects the human eye. Adherence to these limits guarantees the final binocular fusion process.

2. ANSI Z87.1: Occupational and Educational Eye Protection

Industrial eye protection mandates rigorous compliance with high mass and high velocity impact protocols to shield workers from ocular trauma (ANSI, 2020b). Polycarbonate and Trivex substrates must withstand kinetic forces without fracturing or dislodging from the frame. Safety standards reduce the potential for sudden permanent blindness or injury in hazardous industrial environments.

3. ANSI Z80.5: Requirements for Ophthalmic Frames

Structural integrity of the chassis surrounding the optical center demands strict metallurgical and polymeric baseline testing to avoid spontaneous degradation (ANSI, 2010). Flammability and biocompatibility tests protect the wearer from dermatological toxicity and severe thermal hazards.

Chapter Nineteen

Fitting the frame

IN the early eighties, in my first semester of an Ophthalmic Dispensing degree I had an experienced and knowledgeable professor, Art O'Neill. He was tasked with teaching all us rookies how to dispense eyeglasses to patients. Day one he described the dramatic differences in skill levels in the world of opticianry. He stated that fitting glasses is an art. Many become proficient at this art but just as many never really get it. Why? I am convinced that the plethora of bad eyeglass adjustments are the result of opticians that just do not place value on well fitted eyewear. Some corporate leadership is guilty of not encouraging this skill because it never shows up on a spreadsheet in a fashion they can quantify. Remakes show up on a spreadsheet. The remakes that are avoided by a proper adjustment (or readjusted by a more skilled practitioner)do not.

The classic example it the progressive lens wearer that complains of blurry distance vision. Instead of raising the nose pads (to lower the frame) or just spreading the nose pads(to lower the frame) the seg height is remeasured on the poorly fit frame and new lenses reordered. Complete waste of time and resources and extra inconvenience for the patient.

Time and time again I have seen opticians hand the frame to the patient and ask them how it feels. Nine time out of ten if it was bench aligned it will feel fine. Then another optician has to actually adjust the frame three days later when the patient is back with complaints.

Nothing causes more bad eyeglass adjustments than a lazy optician.

There is no book that will really teach you the nuances of fitting eyeglasses. Practice and caring deeply about the final result will be your best teacher. Caring that another optician will not have to fix your work next week is an excellent teacher. Fitting is an art and art demands that you start by giving a damn. No optician will earn a commission for fitting glasses well but you will earn self-respect and the respect of your peers.

It all starts with a bench alignment. Bad bench alignment equals bad adjustment experience 100% of the time. Start with the basics and grow with experience. Bench alignment is simply a frame that has four point touch when place UPSIDE DOWN on the table. Place the frame right side up and if the temple bends are different the frame will look crooked. Bench align upside down. The temples are exactly 90° to the frame front. The frame front has a gentle curve (face form) and there is 8 to 12 degree of pantoscopic tilt (lens/frame front tilt). Use the following graphic as a beginning guide. Practice on old frames.

Bench Alignment Starts Here

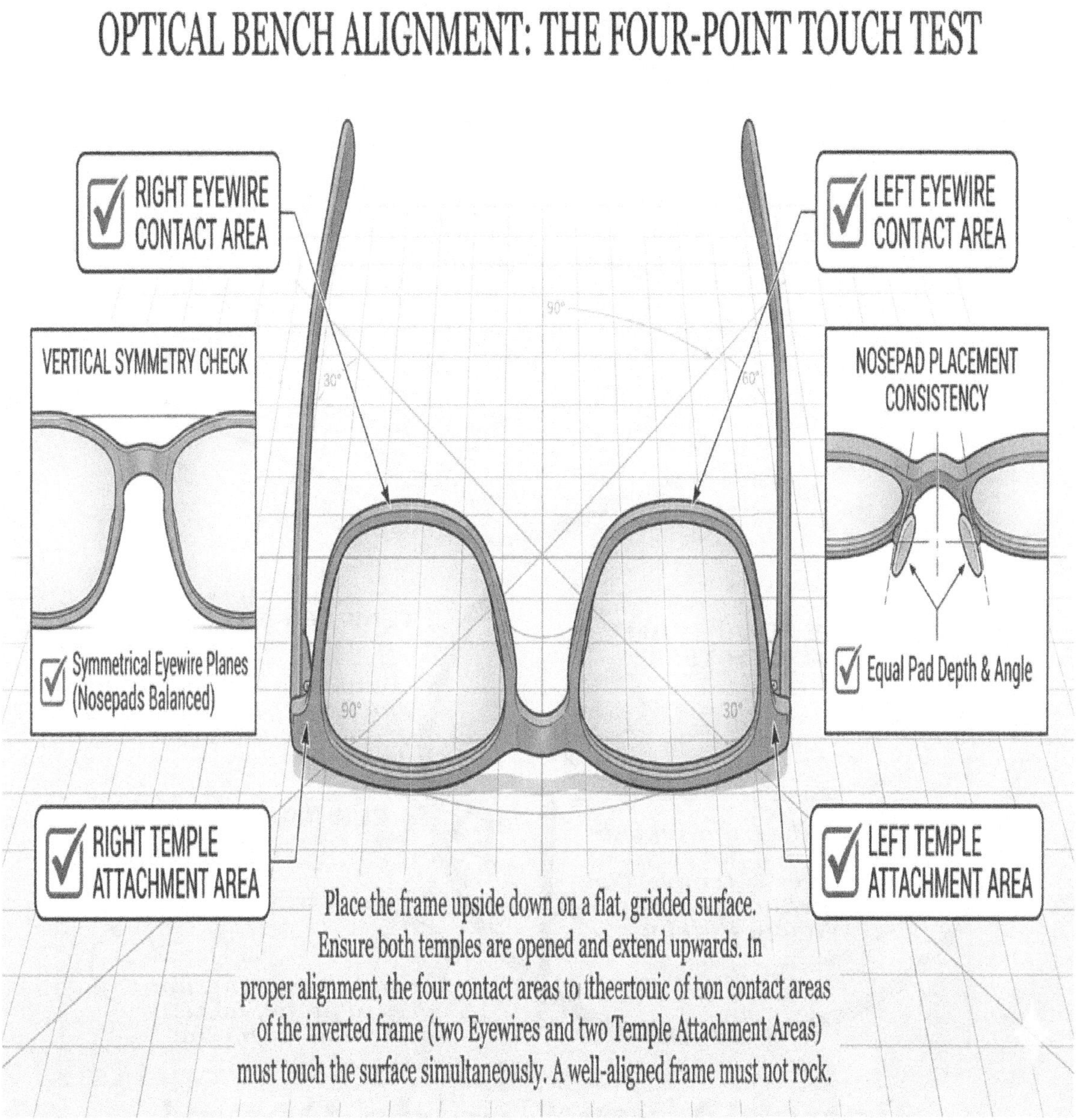

Bench Alignment Principles

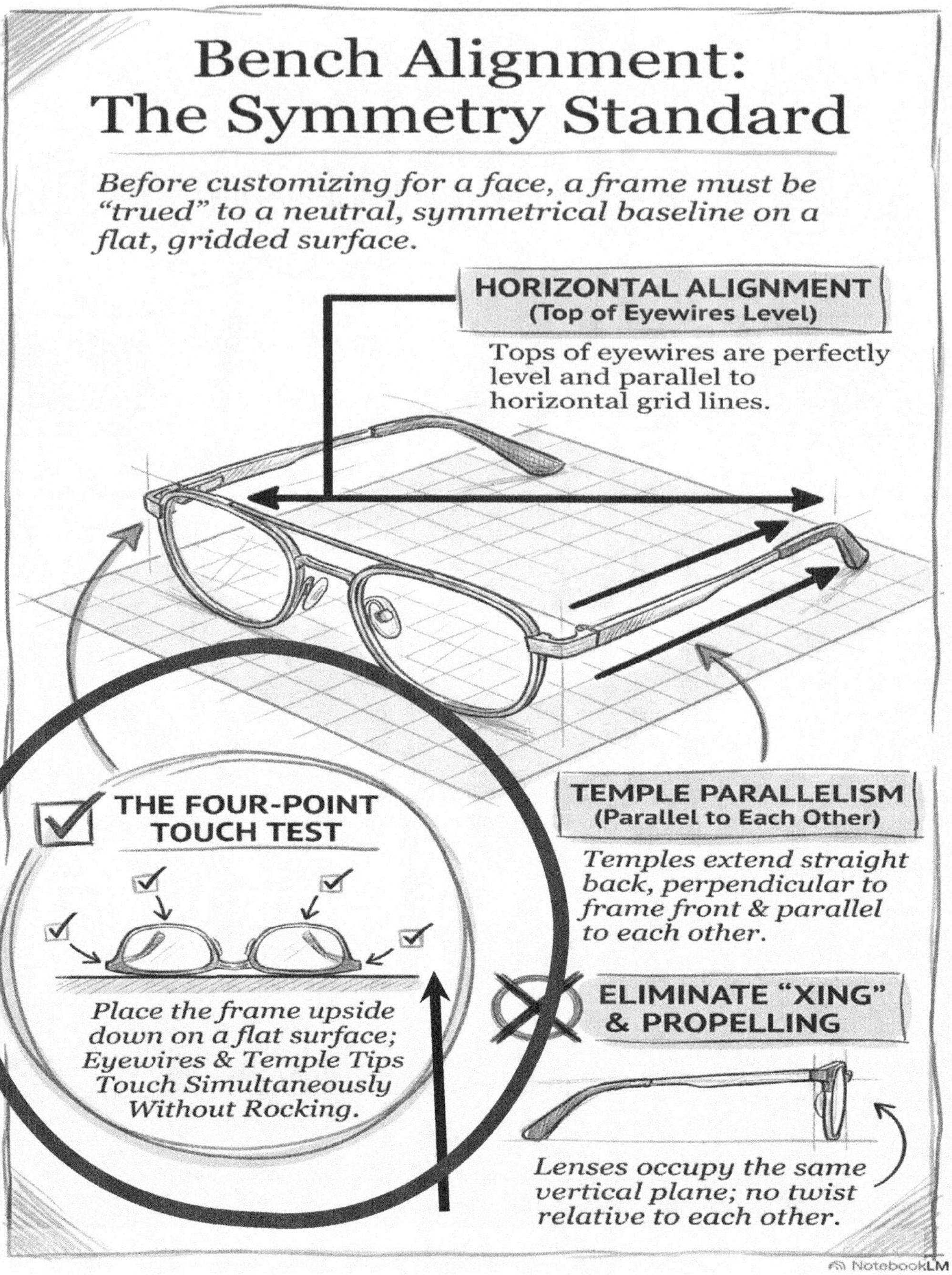

Bench Align Upside-Down!

Bench Alignment Principles

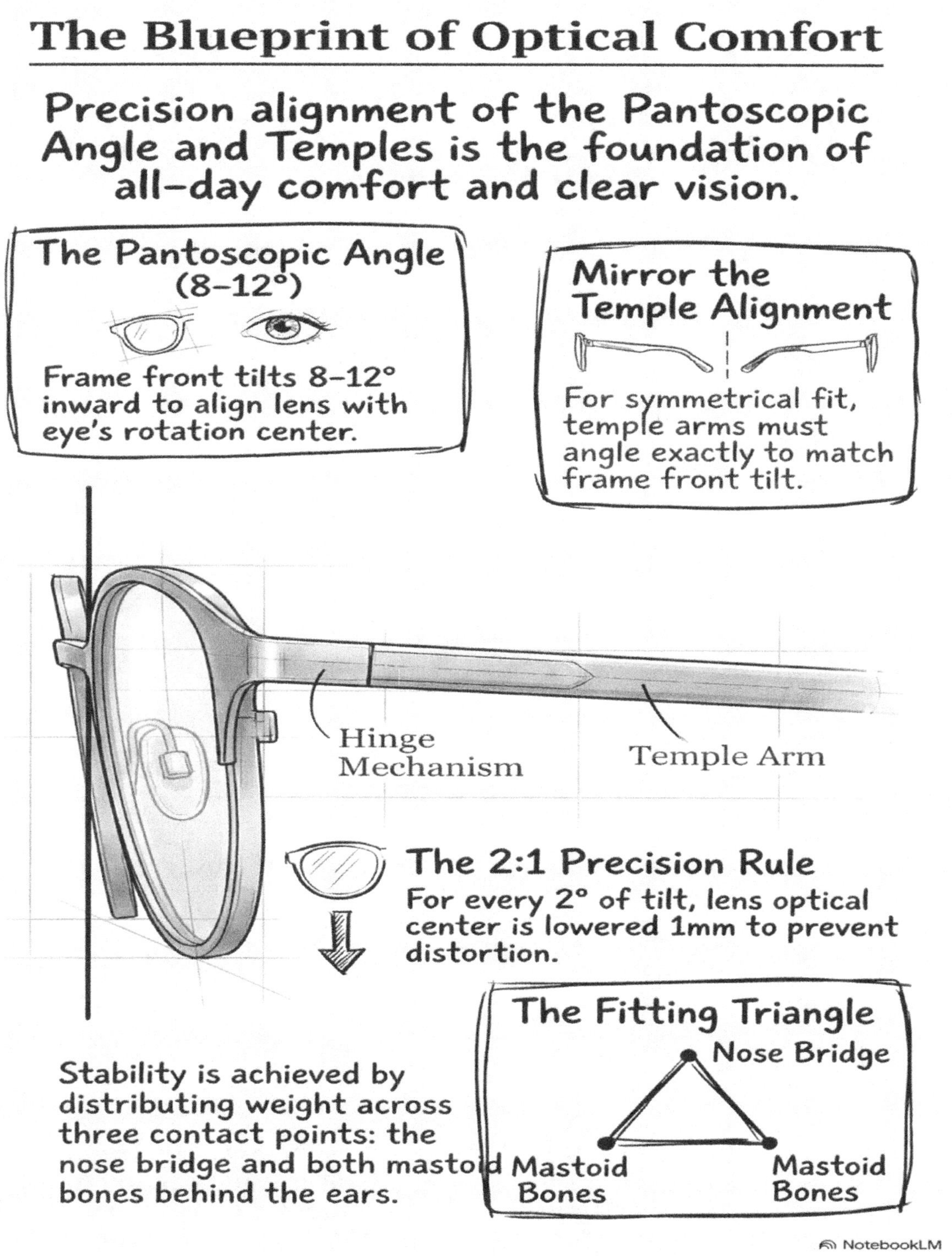

Bench Alignment Principles

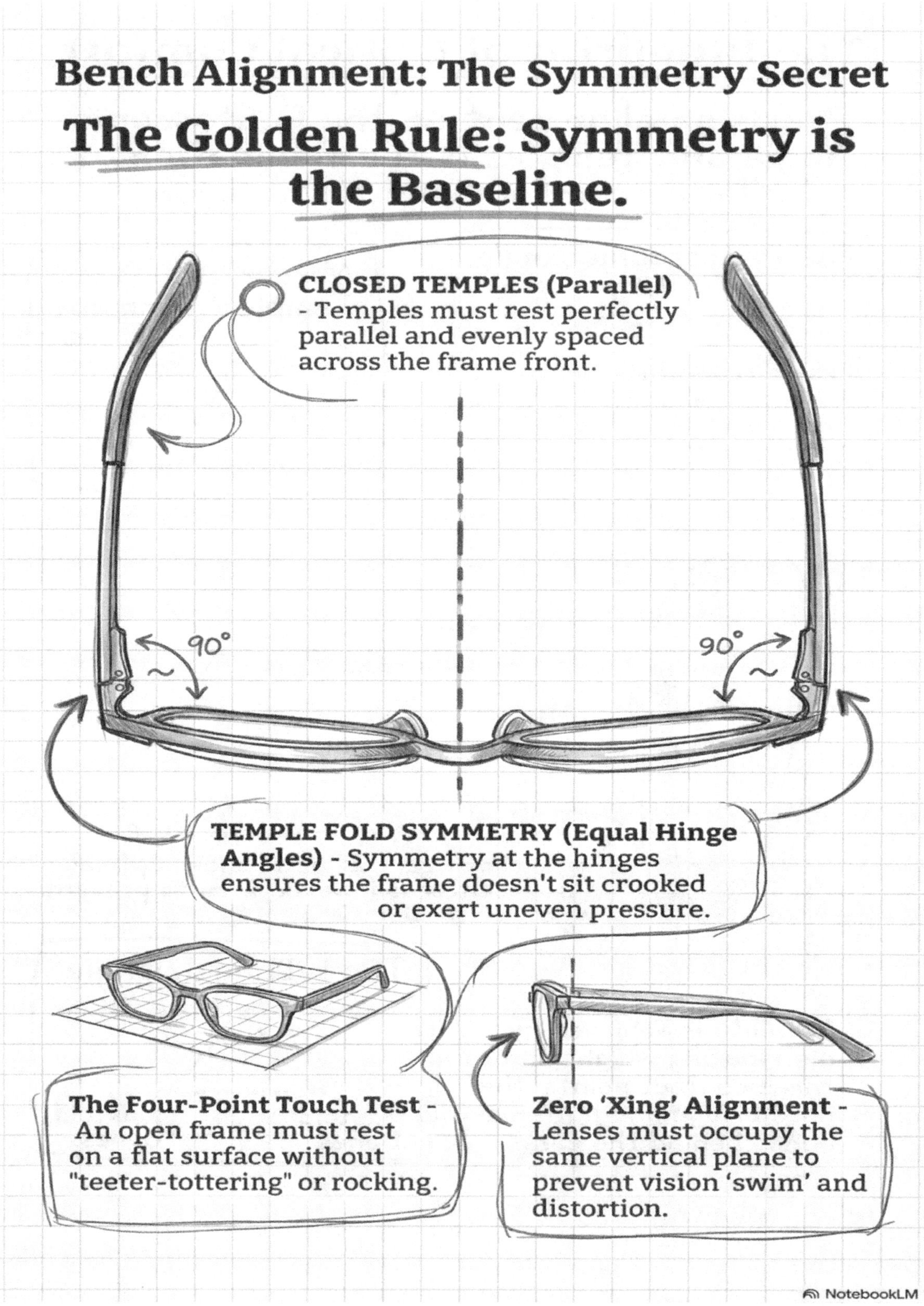

Once you have the frame bench aligned and you are facing the patient place the frame on the patient. Gently slide it over the ears with your pointing finger placed on the bow. I use that finger to physically locate where I want to start the temple bend. I slide my finger tip to the top of their ear and keep it there as I remove the frame from the patients face. That is where I adjust the bent to start just barely behind the top of the ear. Just prior to marking that spot with my finger tip I slide that finger down the back of the ear along the temple tip. This is to make a mental note of the shape and side of the patients' "mastoid process"; the bone behind the ear. Some folks mastoid process are prominent and other are subtle. This is the "art" part. Shape that temple tip to the shape behind the ear. That will maximize the comfort and secure fit. The rest is geometry. Is the frame level on the patient? Does the front curve (face form) seem to curved or to flat? Is there "twist" in the bridge (propeller)? Does the frame sit to low or too high. Are the optical centers (that you dotted up with ink) placed in front of the patients pupils? Is the top of a bifocal segment or PAL to high and in their distance vision. Do they have to pick their head up to read with a bifocal (sitting too low). More pantoscopic tilt lowers the optical center. Spreading nose pads lowers the centers. Raising the nose pads lowers the centers. These are the tools of the pro.

These are the things a competent profession automatically and intuitively thinks of during the dispense. These are the things that will avoid the patient returning next week with complaints.

Take pride in the seeing the geometry of a well fitted frame. Experiment at length with your own glasses even if they are just a pair of cheap sunglasses. You will learn by observing and doing and most important, caring about the result.

Pulling it all Together with Practice

THE secret to becoming a certified optician is not the memorization and rote regurgitation of facts. It is far more important to visualize and understand your options. With that in mind you will achieve your goals with practice, followed by more practice. Use the concepts from this book to determine correct answers to the following situation you will encounter in your career as an optician. Calculate answers and fill in blanks, then compare to answer key.

Metric Conversion Practice

1	.4m	=	_____cm
2	.04m	=	_____cm
3	.004m	=	_____mm
4	62mm	=	_____m
5	620mm	=	_____m
6	.053m	=	_____cm
7	.0053m	=	_____mm
8	68mm	=	_____cm
9	68mm	=	_____m
10	82mm	=	_____m
11	82mm	=	_____cm
12	820mm	=	_____m
13	820mm	=	_____cm
14	4.2m	=	_____cm
15	4.2m	=	_____mm

Prism Practice

1	5 BU OD	Is the same as	5_____OS
2	3 BD OS	Is the same as	3_____OD
3	8 BU	Is the same as	4 BU and_____BD
4	8 BI prescribed & OD has 4BI	Then	OS needs _________
5	3 BO OS	Is the same as	3_____OD
6	3 BU OD	Is the same as	3_____OS
7	10 BU OD	Would split to	5 BU OD and __________

Focal Length Practice

Fill in missing data.

	Focal Length in Meters	Diopters	Focus Real or Virtual?
1	.1m		
2		-10.00	
3		-1.50	
4	.667m		
5	-.25m		
6		+0.25	
7	-.4m		
8	-.2m		
9		+4.00	
10		+1.00	

Lens Power Formula Practice

Fill in missing data.

	Front Surface Curve	Back Surface Curve	Nominal Power
1	+12.00		+8.00
2	+8.00	-4.00	
3	+5.50	-3.50	
4	+6.25		-0.75
5	+2.00		-5.50
6	+22.00	-4.00	
7		-11.50	-7.25
8		-3.00	+3.00

Vertex Distance Power Change Practice

Rrx = Rx/ (1 +d x Rx) (away is negative, closer is positive for d)

	Prescribed Power	Distance Change	Resultant Rx
1	-10.00	5mm closer	
2	-10.00	5mm away	
3	-7.00	7mm closer	
4	-7.00	6mm away	
5	+7.00	4mm closer	
6	+7.00	8mm away	
7	+10.00	6mm closer	
8	+10.00	3mm away	

Optical Cross Practice

	Rx	Sphere Meridian	Cyl Meridian	Rule -With/ Against Oblique
1	-1.00-1.00x90	90		
2	-2.00-1.00x70	70		
3	-3.00-1.00x50	50		
4	-4.00-1.00x30	30		
5	-5.00-1.00x10	10		
6	-6.00-1.00x170	170		
7	-7.00-1.00x150	150		
8	-8.00-1.00x130	130		
9	-9.00-1.00x110	110		

Eye Muscle Practice -Right Eye

	Muscle
1	Superior Rectus
2	Medial Rectus
3	Inferior Rectus
4	Lateral Rectus
5	Superior Oblique
6	Inferior Oblique

Muscle Number	Muscle Movement - OD
	Downward
	Down and Out
	Upward
	Up and Out
	Outward
	Inward

N Value Formula Practice

N = 299,792,485/speed of light in a material

	Speed of Light (m/s)	Speed in Material (m/s)	n-value
1	299,792,485	172,294,512	
2	299,792,485	176,348,521	
3	299,792,485	189,024,265	
4	299,792,485	194,670,445	
5	299,792,485	196,843,391	
6	299,792,485	200,128,495	
7	299,792,485	299,792,485	

The Rx, Plus and Minus Cylinder Practice

Convert the following to plus cylinder

#1	Sphere	Cylinder	Axis	Prism	Base	Add
OD	-1.00	-1.00	180			
OS	-2.00	-0.50	175			
	Sphere	**Cylinder**	**Axis**	**Prism**	**Base**	**Add**
OD						
OS						

#2	Sphere	Cylinder	Axis	Prism	Base	Add
OD	+3.25	-4.25	135			
OS	+3.00	-3.00	160			
	Sphere	**Cylinder**	**Axis**	**Prism**	**Base**	**Add**
OD						
OS						

#3	Sphere	Cylinder	Axis	Prism	Base	Add
OD	+1.50	-1.25	75			
OS	-1.00	-1.25	100			
	Sphere	**Cylinder**	**Axis**	**Prism**	**Base**	**Add**
OD						
OS						

#4	Sphere	Cylinder	Axis	Prism	Base	Add
OD	-14.25	-2.25	90			
OS	-13.00	-4.00	110			
	Sphere	**Cylinder**	**Axis**	**Prism**	**Base**	**Add**
OD						
OS						

Convert the following to minus cylinder

#5	Sphere	Cylinder	Axis	Prism	Base	Add
OD	+3.00	+3.00	10			
OS	+2.50	+2.50	170			
	Sphere	**Cylinder**	**Axis**	**Prism**	**Base**	**Add**
OD						
OS						

#6	Sphere	Cylinder	Axis	Prism	Base	Add
OD	-5.75	+1.00	180			
OS	-5.50	+1.00	10			
	Sphere	**Cylinder**	**Axis**	**Prism**	**Base**	**Add**
OD						
OS						

#7	Sphere	Cylinder	Axis	Prism	Base	Add
OD	-1.00	+1.00	35			
OS	-1.25	+1.25	40			
	Sphere	Cylinder	Axis	Prism	Base	Add
OD						
OS						

#8	Sphere	Cylinder	Axis	Prism	Base	Add
OD	0.00	+2.25	175			
OS	-1.25	+1.50	160			
	Sphere	Cylinder	Axis	Prism	Base	Add
OD						
OS						

Calculate the Power on the 180 Practice

Power 180 = Cyl Power x (Cyl degrees to 180/90) + Sphere Power

You are looking for the percentage of the cylinder that needs to be added to the sphere to determine the power on the 180 (or zero) line.

#1	Sphere	Cylinder	Axis
OD	-5.00	-3.00	70
OS	-4.50	-2.50	100
180 Power OD			
180 Power OS			

#2	Sphere	Cylinder	Axis
OD	+5.00	-3.00	70
OS	+4.50	-2.50	100
180 Power OD			
180 Power OS			

#3	Sphere	Cylinder	Axis
OD	+2.00	-3.00	75
OS	+1.50	-2.50	88
180 Power OD			
180 Power OS			

#4	Sphere	Cylinder	Axis
OD	-5.00	-1.00	30
OS	-4.00	-2.50	180
180 Power OD			
180 Power OS			

Prentice's Rule Practice #1

Use everything you have learned and visualize the placement of the OC's compared to the patients PD, the centers of the pupil.

Prism = (C)(D)/10   C=decentration mm   D = Diopters

Upon inspecting this patient's glasses, you have discovered that the OC's dot up, horizontally, to 34mm in the right eye and 29mm is the left eye. Determine the amount of unwanted prism and direction of prism induced in each eye (base direction).

Then determine total prism for both eyes together. Refer to the ANSI standards Do these glasses pass inspection as far as horizontal prism is concerned? See Illustration that follows.

#1	Sphere	Cylinder	Axis	
OD	+5.00			
OS	+5.00			
				Prism Found
PD OD	31		OD	
PD OS	32		OS	
			OU	

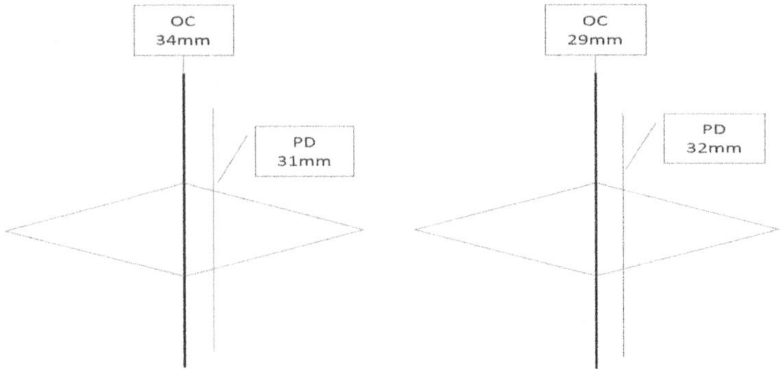

↑ Prentice Rule Practice #1 ↑

Prentice's Rule Practice #2

#2	Sphere	Cylinder	Axis	
OD	-7.25			
OS	-5.00			
				Prism Found
PD OD	31		OD	
PD OS	32		OS	
			OU	

Upon inspecting this patient's glasses, you have discovered that the OC's dot up, horizontally, to 34mm in the right eye and 29mm is the left eye. Determine the amount of unwanted prism and direction of prism induced in each eye (base direction).

Do these glasses pass inspection as far as horizontal prism is concerned?

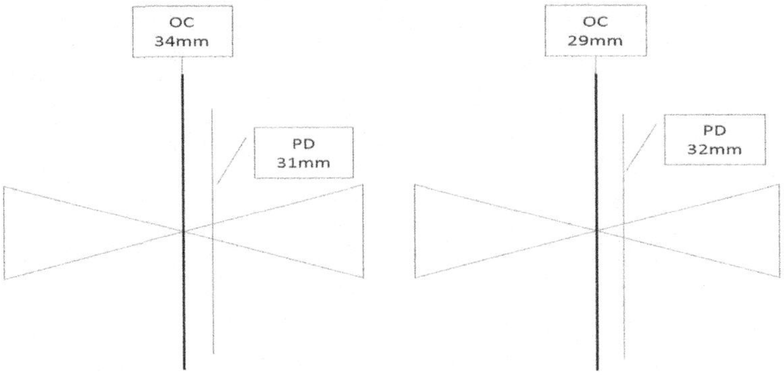

Prentice's Rule Practice #3

180 Power = Cyl Power x (degrees to 180/90) +Sphere

#3	Sphere	Cylinder	Axis	
OD	-5.00	-3.00	45	
OS	-5.00	-3.50	180	
				Prism Found
PD OD	31		OD	
PD OS	32		OS	
			OU	

Upon inspecting this patient's glasses, you have discovered that the OC's dot up, horizontally, to 34mm in the right eye and 29mm is the left eye. Determine the amount of unwanted prism and di- rection of prism induced in each eye (base direction).

Do these glasses pass inspection as far as horizontal prism is concerned?

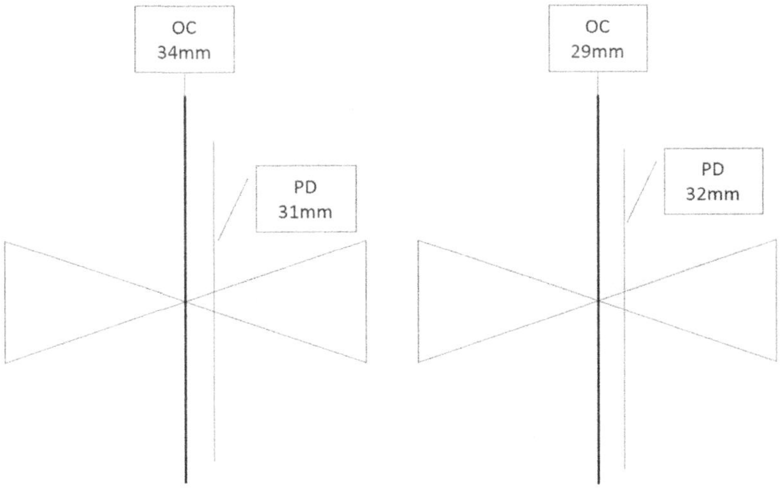

Prentice's Rule Practice #4

#4	Sphere	Cylinder	Axis	
OD	-5.00	-3.00	45	
OS	-5.00	-3.50	180	
				Vertical Prism Found
PD OD	31		OD	
PD OS	32		OS	
			OU	

Upon inspecting this patient's glasses, you have discovered that the OC's dot up, horizontally, to 31mm in the right eye and 32mm is the left eye. However, the right eye OC is on the B line. The left eye OC is 4 mm above the B line. Determine the amount of unwanted prism and direction of prism induced in each eye (base direction).

Do these glasses pass inspection as far as vertical prism is concerned?

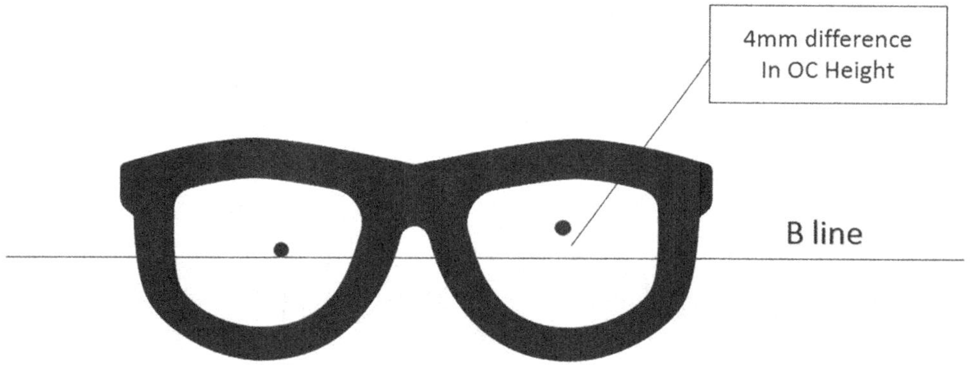

PRACTICE ANSWERS

Metric Conversion Practice Answers

1	.4m	=	40 cm
2	.04m	=	4 cm
3	.004m	=	4 mm
4	62mm	=	.062 m
5	620mm	=	.62 m
6	.053m	=	53 cm
7	.0053m	=	5.3 mm
8	68mm	=	6.8 cm
9	68mm	=	.0068 m
10	82mm	=	.0082 m
11	82mm	=	8.2 cm
12	820mm	=	.82 m
13	820mm	=	82 cm
14	4.2m	=	420 cm
15	4.2m	=	4200 mm

Prism Practice Answers

1	5 BU OD	Is the same as	5 BD OS
2	3 BD OS	Is the same as	3 BU OD
3	8 BU	Is the same as	4 BU and 4 BD
4	8 BI prescribed & OD has 4BI	Then	OS needs 4 BI

5	3 BO OS	Is the same as	3 BO OD
6	3 BU OD	Is the same as	3 BD OS
7	10 BU OD	Would split to	5 BU OD 5 BD OS

Focal Length Practice Answers

	Focal Length in Meters	Diopters	Focus Real or Virtual?
1	.1m	+10.00	R
2	-.1m	-10.00	V
3	-.667m	-1.50	V
4	.667m	+1.50	R
5	-.25m	-4.00	V
6	4m	+0.25	R
7	-.4m	-2.50	V
8	-.2m	-5.00	V
9	.25m	+4.00	R
10	1.0m	+1.00	R

Lens Power Formula Answers

	Front Surface Curve	Back Surface Curve	Nominal Power
1	+12.00	-4.00	+8.00
2	+8.00	-4.00	+4.00
3	+5.50	-3.50	+2.50
4	+6.25	-7.00	-0.75
5	+2.00	-7.50	-5.50
6	+22.00	-4.00	+18.00
7	+4.25	-11.50	-7.25
8	+6.00	-3.00	+3.00

Vertex Distance Power Change Practice Answers

Rrx = Rx/(1 +d x Rx) (away is negative, closer is positive for d)

	Prescribed Power	Distance Change	Resultant Rx
1	-10.00	5mm closer	-10.53
2	-10.00	5mm away	-9.52
3	-7.00	7mm closer	-7.37
4	-7.00	6mm away	-6.73
5	+7.00	4mm closer	+6.80
6	+7.00	8mm away	+7.45
7	+10.00	6mm closer	+9.43
8	+10.00	3mm away	+10.31

Optical Cross Practice Answers

	Rx	Sphere Meridian	Cyl Meridian	Rule -With/ Against Oblique
1	-1.00-1.00x90	90	180	Against
2	-2.00-1.00x70	70	160	Against
3	-3.00-1.00x50	50	140	Oblique
4	-4.00-1.00x30	30	120	Oblique
5	-5.00-1.00x10	10	100	With
6	-6.00-1.00x170	170	80	With
7	-7.00-1.00x150	150	60	Oblique
8	-8.00-1.00x130	130	40	Oblique
9	-9.00-1.00x110	110	20	Against

Eye Muscle Practice Answers

Right Eye

	Muscle
1	Superior Rectus
2	Medial Rectus
3	Inferior Rectus
4	Lateral Rectus
5	Superior Oblique
6	Inferior Oblique

Muscle Number	Muscle Movement - OD
3	Downward
5	Down and Out
1	Upward
6	Up and Out
4	Outward
2	Inward

N Value Formula Practice Answers

N = 299,792,485/speed of light in a material

	Speed of Light (m/s)	Speed in Material (m/s)	n-value
1	299,792,485	172,294,512	1.74
2	299,792,485	176,348,521	1.70
3	299,792,485	189,024,265	1.586
4	299,792,485	194,670,445	1.54
5	299,792,485	196,843,391	1.523
6	299,792,485	200,128,495	1.498
7	299,792,485	299,792,485	1.00

The Rx, Plus and Minus Cylinder Practice Answers

Convert the following to plus cylinder

#1	Sphere	Cylinder	Axis	Prism	Base	Add
OD	-1.00	-1.00	180			
OS	-2.00	-0.50	175			
	Sphere	**Cylinder**	**Axis**	**Prism**	**Base**	**Add**
OD	-2.00	+1.00	90			
OS	-2.50	+0.50	85			

#2	Sphere	Cylinder	Axis	Prism	Base	Add
OD	+3.25	-4.25	135			
OS	+3.00	-3.00	160			
	Sphere	**Cylinder**	**Axis**	**Prism**	**Base**	**Add**
OD	-1.00	+4.25	45			
OS	0.00	+3.00	70			

#3	Sphere	Cylinder	Axis	Prism	Base	Add
OD	+1.50	-1.25	75			
OS	-1.00	-1.25	100			
	Sphere	**Cylinder**	**Axis**	**Prism**	**Base**	**Add**
OD	+0.25	+1.25	165			
OS	-2.25	+1.25	10			

#4	Sphere	Cylinder	Axis	Prism	Base	Add
OD	-14.25	-2.25	90			
OS	-13.00	-4.00	110			
	Sphere	**Cylinder**	**Axis**	**Prism**	**Base**	**Add**
OD	-16.50	+2.25	180			
OS	-17.00	+4.00	20			

Convert the following to minus cylinder

#5	Sphere	Cylinder	Axis	Prism	Base	Add
OD	+3.00	+3.00	10			
OS	+2.50	+2.50	170			
	Sphere	**Cylinder**	**Axis**	**Prism**	**Base**	**Add**
OD	+6.00	-3.00	100			
OS	+5.00	-2.50	80			

#6	Sphere	Cylinder	Axis	Prism	Base	Add
OD	-5.75	+1.00	180			
OS	-5.50	+1.00	10			
	Sphere	**Cylinder**	**Axis**	**Prism**	**Base**	**Add**
OD	-4.75	-1.00	90			
OS	-4.50	-1.00	100			

#7	Sphere	Cylinder	Axis	Prism	Base	Add
OD	-1.00	+1.00	35			
OS	-1.25	+1.25	40			
	Sphere	**Cylinder**	**Axis**	**Prism**	**Base**	**Add**
OD	0.00	-1.00	125			
OS	0.00	-1.25	130			

#8	Sphere	Cylinder	Axis	Prism	Base	Add
OD	0.00	+2.25	175			
OS	-1.25	+1.50	160			
	Sphere	**Cylinder**	**Axis**	**Prism**	**Base**	**Add**
OD	+2.25	-2.25	85			
OS	+0.25	-1.50	70			

Calculate the Power on the 180 Practice Answers

Power 180 = Cyl Power x (Cyl degrees to 180/90) + Sphere Power You are looking for the percentage of the cylinder that needs to be added to the sphere to determine the power on the 180 (or zero) line.

#1	Sphere	Cylinder	Axis
OD	-5.00	-3.00	70
OS	-4.50	-2.50	100
180 Power OD	-7.33		
180 Power OS	-6.72		

#2	Sphere	Cylinder	Axis
OD	+5.00	-3.00	70
OS	+4.50	-2.50	100
180 Power OD	+2.67		
180 Power OS	+2.28		

#3	Sphere	Cylinder	Axis
OD	+2.00	-3.00	75
OS	+1.50	-2.50	88
180 Power OD	-0.50		
180 Power OS	-0.94		

#4	Sphere	Cylinder	Axis
OD	-5.00	-1.00	30
OS	-4.00	-2.50	180
180 Power OD	-5.33		
180 Power OS	-4.00		

Prentice's Rule Practice Answer #1

Use everything you have learned and visualize the placement of the OC's compared to the patients PD, the centers of the pupil.

Prism = (C)(F)/10 C=decentration mm F = Focal in Diopters

#1	Sphere	Cylinder	Axis	
OD	+5.00			
OS	+5.00			
				Prism Found
PD OD	31		OD	1.50 BO
PD OS	32		OS	1.50 BI
			OU	0.00

Upon inspecting this patient's glasses, you have discovered that the OC's dot up, horizontally, to 34mm in the right eye and 29mm is the left eye. Determine the amount of unwanted prism and direction of prism induced in each eye (base direction).

Then determine total prism for both eyes together. Refer to the ANSI standards.

Do these glasses pass inspection as far as horizontal prism is concerned? (YES)

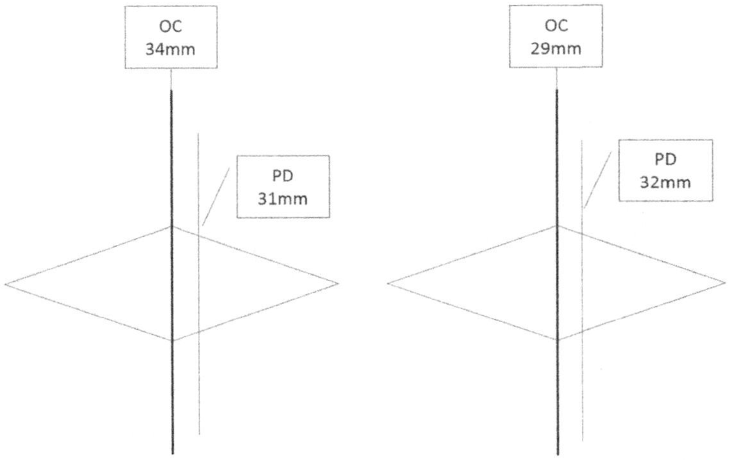

Prentice's Rule Practice Answer #2

#2	Sphere	Cylinder	Axis	
OD	-7.25			
OS	-5.00			
				Prism Found
PD OD	31		OD	2.18 BI
PD OS	32		OS	1.50 BO
			OU	0.68 BI

Upon inspecting this patient's glasses, you have discovered that the OC's dot up, horizontally, to 34mm in the right eye and 29mm is the left eye. Determine the amount of unwanted prism and direction of prism induced in each eye (base direction).

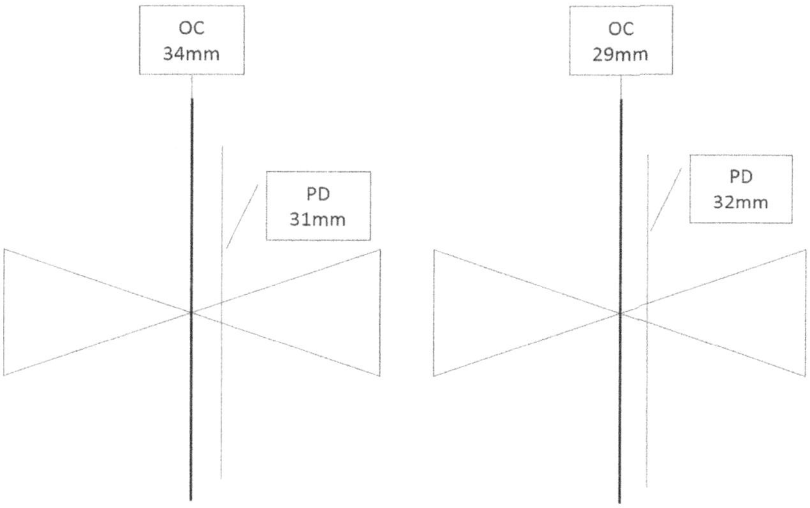

Do these glasses pass inspection as far as horizontal prism is concerned? (NO)

Prentice's Rule Practice Answer #3

180 Power = Cyl Power x (degrees to 180/90) +Sphere

#3	Sphere	Cylinder	Axis	
OD	-5.00	-3.00	45	(-6.50 on 180)
OS	-5.00	-3.50	180	(-5.00 on 180)
				Prism Found
PD OD	31		OD	1.95 BI
PD OS	32		OS	1.50 BO
			OU	0.45 BI

Upon inspecting this patient's glasses, you have discovered that the OC's dot up, horizontally, to 34mm in the right eye and 29mm is the left eye. Determine the amount of unwanted prism and direction of prism induced in each eye (base direction).

Do these glasses pass inspection as far as horizontal prism is concerned? (YES)

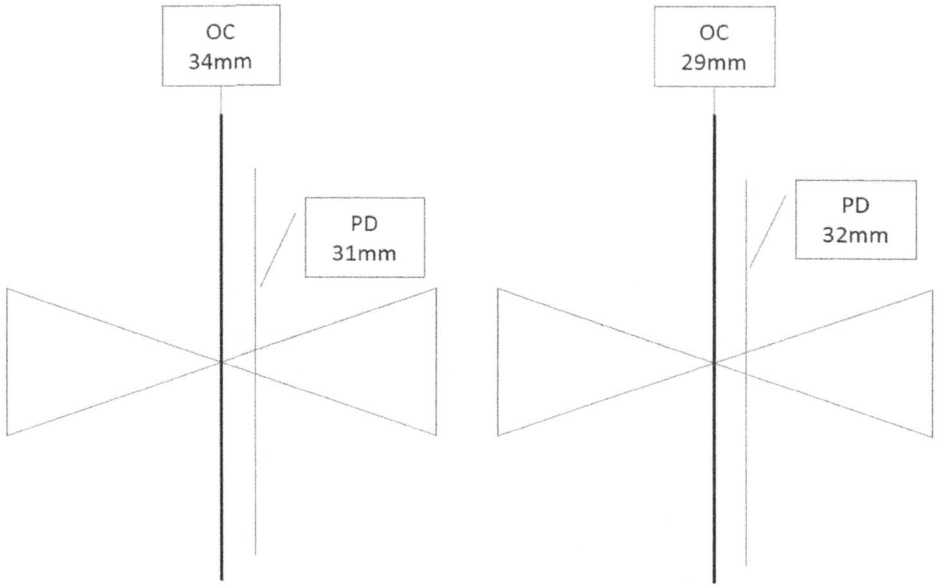

Prentice's Rule Practice Answer#4

#4	Sphere	Cylinder	Axis	
OD	-5.00	-3.00	45	
OS	-5.00	-3.50	180	
				Vertical Prism Found
PD OD	31		OD	0.00
PD OS	32		OS	2.00 BD
			OU	2.00 Vertical Imbalance

Upon inspecting this patient's glasses, you have discovered that the OC's dot up, horizontally, to 31mm in the right eye and 32mm is the left eye. However, the right eye OC is on the B line. The left eye OC is 4 mm above the B line. Determine the amount of unwanted prism and direction of prism induced in each eye (base direction).

Do these glasses pass inspection as far as vertical prism is concerned? (NO)

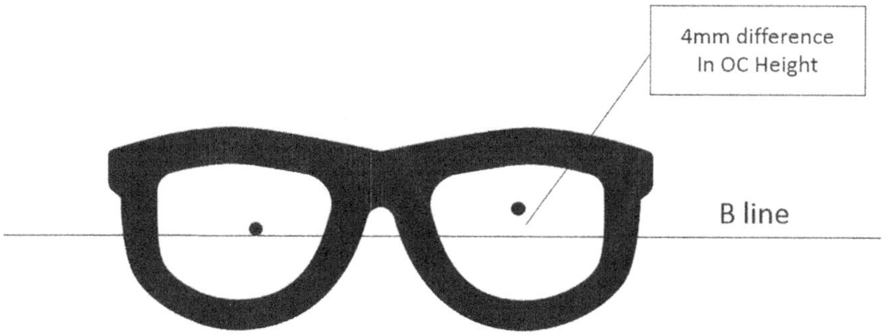

These examples were single vision lenses and illustrate prism in and out of tolerance. Remember, with a bifocal, and in particular, a PAL, you would never pass inspection with these discrepancies. **A** monocular PD such as that would make the distance between the reading areas unacceptable.

ABO TOPICS EXAM

Chapter One Practice Exam

[#001] What is the most common critical measurement an optician will make on a day-to-day basis?

a) The speed of light
b) MBS
c) PD
d) Decentration

[#002] Which unit is a metric linear unit?

a) Liter
b) Gram.
c) Metric Tonne
d) Millimeter

[#003] How many millimeters in an inch?

a) 254
b) .254
c) 25.4
d) 2.54

[#004] The vast majority of patient PDs will be:

a) Between 55 to 75 millimeters
b) Between 5.5 to 7.5 millimeters
c) Between 55 to 75 centimeters
d) Between .5 to .7 meters

[#005] The base unit of linear metric measurements is a:

a) Kilometer
b) Nanometer
c) Meter
d) Millimeter

[#006] Which is the smallest linear metric unit on this list?
a) Kilometer
b) Nanometer
c) Centimeter
d) Meter

[#007] Which is the smallest linear metric unit on this list?
a) Centimeter
b) Meter
c) Kilometer
d) 1/10th of an inch

[#008] Which is the largest linear metric unit on this list?
a) Milliliter
b) Nanometer
c) Nanoliter
d) Microgram

[#009] Which linear metric unit is closest to 4/100th of an inch?
a) One kilometer
b) One meter
c) One millimeter
d) One nanometer

[#010] The closest metric unit in length to a yard is
a) One kilometer
b) One meter
c) One millimeter
d) One nanometer

[#011] How many millimeters in a centimeter?
a) .01
b) 0.1
c) 1.0
d) 10

[#012] How many centimeters in a meter?

a) 0.1
b) 1.0
c) 10.0
d) 100

[#013] How many meters in a kilometer?

a) 1.0
b) 10.0
c) 100
d) 1000

[#014] The patient's PD = 60mm. What is that in meters?

a) .006 meters
b) .06 meters
c) .6 meters
d) 6 meters

[#015] 64 millimeters equals:

a) 6.4 cm and .064 m
b) .64 cm and .064 m
c) .064 cm and .64 m
d) None of the above

[#016] 60 mm divided by 10 converts the mm to

a) .06 cm
b) .06 km
c) 6 cm
d) 6 nm

[#017] .06 m multiplied by 1000 converts the m to:

a) 600 mm
b) 60 mm
c) 60 cm
d) 6 km

[#018] You are seated across from an optical engineer who tells you the distance between his eyes is .059 meters (trust me, this will happen someday). If his number is accurate, how many millimeters would that equal?

a) 5.9
b) .059
c) 59
d) 590

[#019] 25.4 mm equals:

a) One foot
b) One yard
c) One inch
d) One acre

[#020] 39.4 inches equals:

a) One foot
b) One yard
c) One inch
d) One meter

[#021] How many nanometers in a meter?

a) 1,000
b) 1,000,000
c) 1,000,000,000
d) 1,000,000,000,000

[#022] How many millimeters in a meter

a) 1,000
b) 1,000,000
c) 1,000,000,000
d) 1,000,000,000,000

[#023] How many meters in a nanometer?

a) .0001
b) .0000001
c) .0000000001
d) .000000000001

[#024] Which statement is the most accurate?

a) A km is smaller than a meter and a mm is smaller than a nanometer

b) A km is larger than a meter and a cm is smaller than a nanometer

c) It is easier to see an object that is 10 nanometers in size than an object that is 1/10th of a centimeter

d) A meter is one billion times longer than a nanometer

[#025] 0.50 m is equivalent to:

a) 50 mm and .05 cm

b) 500 mm and 50 cm

c) .05 cm and 0.50 mm

d) 5000 mm and 500 cm

Chapter Two Practice Exam

[#026] Light, in a vacuum, travels at:
a) Nearly 300 km per second or 186 miles per second
b) Nearly 3000 km per second or 1,860 miles per second
c) Nearly 30000 km per second or 18,600 miles per second
d) Nearly 300000 km per second or 186,000 miles per second

[#027] When light enters a clear medium such as glass it will:
a) Mostly reflect off
b) Stop
c) Slow down
d) Speed up

[#028] White light is composed of:
a) Only the longer wavelengths of the visual spectrum
b) Only the shorter wavelengths of the visual spectrum
c) All the wavelengths of the visual spectrum
d) All the elements of the electromagnetic spectrum

[#029] Wavelength is:
a) The distance between the crests of waves
b) The length of the series of waves
c) Always very short
d) Only received by televisions and radios

[#030] One billionth of a meter is a:
a) Decameter
b) Centimeter
c) Micron
d) Nanometer

[#031] The electromagnetic spectrum includes
a) Gamma rays and X-rays
b) UV & infrared and microwaves
c) Radio waves and the visual spectrum
d) All of the above

[#032] Which statement is correct?

a) All known wavelengths are contained within the visual spectrum

b) UV rays have a longer wavelength than infrared rays

c) Microwaves have a shorter wavelength than UV rays

d) The visual spectrum is near the center of the electromagnetic spectrum

[#033] Blue light waves:

a) Have a longer wavelength than red light waves

b) Have a shorter wavelength than red light waves

c) Have a shorter wavelength than UV waves

d) Have a shorter wavelength than X-ray waves

[#034] The visual spectrum contains:

a) Only blue, yellow, orange and red

b) Many colors and UV rays

c) All the colors of the rainbow

d) Primarily red, yellow and blue

[#035] The wavelengths of the visual spectrum are:

a) Too short to accurately measure

b) Usually measured in millimeters

c) From about 400 nm to 750 nm

d) Exactly 580 nm long

[#036] The wavelengths of the electromagnetic spectrum that tan your skin are:

a) Infrared

b) Yellow light rays

c) Ultraviolet

d) Upper electromagnetic spectrum

[#037] Select the answer that correctly states the order of colors from shorter wavelengths to longer ones:

a) Blue, green, yellow, orange, red

b) Red, orange, yellow, green, blue

c) Blue, orange, yellow, green, red

d) None of the above

[#038] If the visual spectrum is from 400 to 750 nanometers, what is the median or central point of that range?
a) 600 nm
b) 550 nm
c) 650 nm
d) 575 nm

[#039] What is the wavelength of yellow (Sodium D-line)?
a) 550 nm
b) 575 nm
c) 580 nm
d) 600 nm

[#040] Which color is the human eye able to detect the most easily?
a) Blue
b) Green
c) Orange
d) Yellow

[#041] What can be added to ophthalmic lenses to reduce the wavelengths emitted by computer or cell phone screens?
a) Infrared blocker
b) UV blocker
c) Brown tinting
d) Blue blocker

[#042] The color separation that occurs when a rainbow is formed is dependent on:
a) Colors that are all the same wavelength
b) Colors with the same energy levels
c) The differing wavelengths and energies of the colors
d) The angle the sun forms with the axis of the earth

[#043] When diagramming the path of a light ray to help determine the effect of differing lenses and prisms, you should consider a beam of light as:
a) Light traveling in all directions
b) A single ray diagrammed as an arrow
c) Diverging
d) Converging

[#044] The source of a light ray can be known as a:
a) Bulb
b) Pinhole
c) Source of wavelength emanation
d) Point of origin

[#045] When light rays from multiple light sources are redirected by a lens to a single point, the rays are considered to have:
a) Converged
b) Diverged
c) Become parallel
d) Scattered

[#046] Which statement is NOT correct:
a) Green wavelengths fall between blue and yellow wavelengths
b) Red wavelengths are the longest in the visual spectrum
c) Green wavelengths fall between yellow and red wavelengths
d) Yellow is in the center of the visual spectrum

Chapter Three Practice Exam

[#047] A plane (or plano) mirror can also be called a:
- a) 747 rearview mirrors
- b) Incident ray mirror
- c) Normal mirror
- d) Flat mirror

[#048] If a light ray strikes a plano mirror at 60 degrees it shall be reflected at:
- a) $180^\circ - 60^\circ = 120^\circ$
- b) The angle of the normal
- c) 60°
- d) 90°

[#049] If the angle of reflection from a plano mirror is 179 degrees, then the incident ray must have approached:
- a) From multiple points of origin
- b) In a fashion that included both convergence and divergence
- c) At an equivalent angle
- d) From a light source in motion

[#050] The dotted lines on this ray tracing __________ to a point forming a ________ __________:
- a) Diverge, image point
- b) Scatter, source plane
- c) Converge, real image
- d) Converge, virtual image

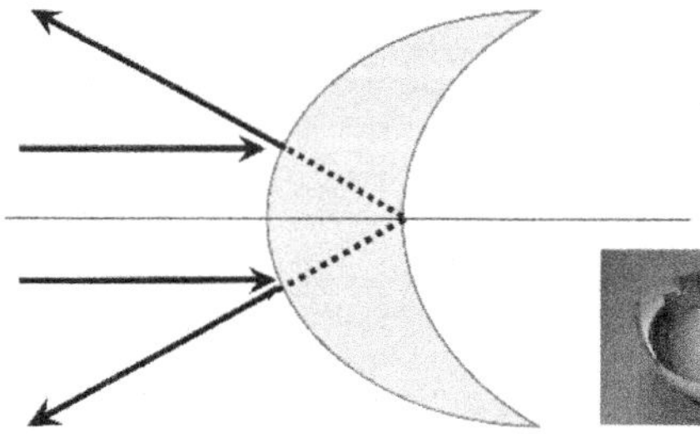

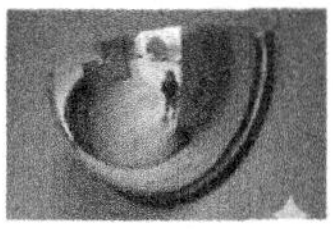

[#051] A mirror that reflects light as indicated by the diagram in question #050 is:
- a) A plano mirror
- b) A convex mirror
- c) A concave mirror
- d) A security mirror

[#052] The distance from the front of the mirror in the diagram (in question #050) to the point where the dotted lines converge can be used:
- a) To adjust the positioning of the mirror
- b) To determine the reflectivity of the mirror
- c) To calculate the scattered reflected ray percentage
- d) To calculate the curvature of the mirror

[#053] Regarding mirrors, which statement is **NOT** correct?

a) An image that is formed in front of a mirror is the result of the effect of a concave mirror
b) The reflected rays of a convex mirror provide a larger field of view and the image undergoes minification
c) A reflecting telescope's primary (largest) mirror must diverge light rays to be effective
d) A concave mirror can make the image look upside down.

[#054] Which statement is true?

a) The convex side mirror on your auto typically works by creating a smaller field of view.
b) The concave mirror that is in your headlamps scatter light to provide a wider field of view.
c) The convex side mirror provides a wider field of view while the concave headlamp mirror helps focus your light beams forward.
d) The convex side mirror provides a smaller field of view while the concave headlamp mirror helps scatter your headlamp light into a wider field of view.

Chapter Four Practice Exam

[#055] Which statement about a glass prism is true?

a) Prism reflects the majority of the light that enters it
b) Prism converges the majority of the light that enters it
c) Prism diverges the majority of the light that enters it
d) Prism refracts the majority of the light that enters it

[#056] An imaginary line perpendicular to the surface of a prism is called:

a) The normal
b) The apex
c) The base
d) The refraction

[#057] When light bends, deviates or turns in a prism, it is called:

a) Divergence
b) Reflection
c) Refraction
d) Convergence

[#058] As light entering a prism refracts:

a) It speeds up
b) It maintains a constant speed
c) It comes to a stop
d) It slows down

[#059] As light entering a prism refracts:

a) Colors with longer wavelengths refract more than shorter ones
b) Colors remain at a constant speed relative to each other
c) Colors with shorter wavelengths refract more than colors with longer wavelengths
d) Colors blend with each other

[#060] The "normal" is:

a) A line perpendicular to a surface of the prism
b) The wavelength of central yellow light
c) The regular separation of colors in a prism
d) Parallel rays

[#061] When colors of different wavelengths refract, the separation of colors is called:

a) Distinction
b) Distribution
c) Dispersion
d) Displacement

[#062] When different colors refract differing amounts, based on differing wavelengths, this can be called dispersion or the:

a) Color spectrum
b) Banded shades
c) Palette
d) Rainbow effect

[#063] Which statement is <u>**NOT**</u> correct?

a) When light rays enter a new and denser clear medium, the rays refract towards the normal
b) When light rays enter a new and less dense medium the rays refract away from the normal
c) The normal is a line perpendicular to the surface of the new medium
d) When light rays refract all colors maintain a constant velocity

[#064] Dispersion occurs when the colors of light:

a) Approach a new transparent medium
b) Leaves a clear medium
c) Refract within a denser medium
d) Reflects from a clear medium

[#065] Which statement is **NOT** correct?

a) When you see a rainbow in the sky you know that light, passing from air into water vapor and out again, is the result of refraction and dispersion that creates the rainbow

b) Understanding how light travels when leaving one medium and entering another gives the optician the power to correct refractive errors in a patient's eyes

c) An optician will never have to consider dispersion in the design of a patient's eyewear

d) Red light will refract less than yellow light in a medium

[#066] In the diagram, identify the letter that indicates the normal as light rays transition from a denser medium to a less dense medium.

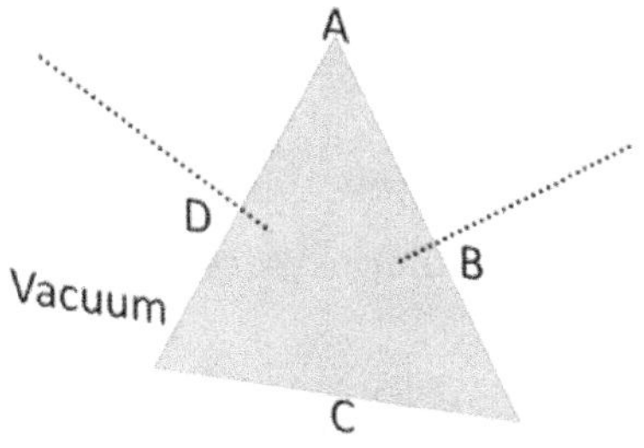

a) B

b) C

c) D

d) A

[#067] In the diagram, which letter represents the base of the prism?

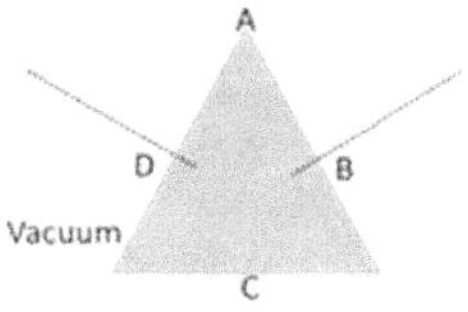

a) A

b) B

c) C

d) D

[#068] In the diagram, which letter represents the apex of the prism?

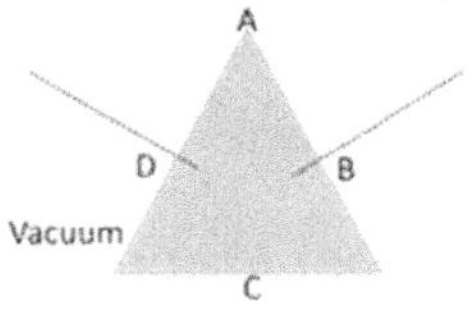

a) A

b) B

c) C

d) D

[#069] In the diagram, identify the letter that indicates the normal as light rays transition from a less dense medium to a denser medium.

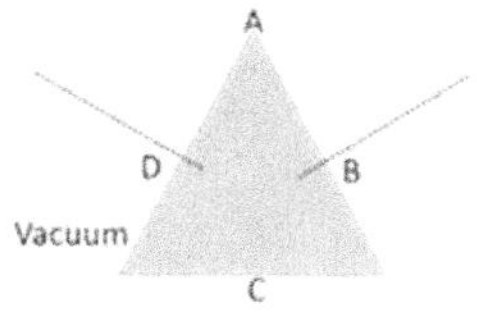

a) A
b) B
c) C
d) D

[#070] Which statement is **<u>NOT</u>** correct?

a) When light passes from air to glass the rays are refracted towards the base (towards the normal)
b) When light passes from glass to air the light is refracted towards the base (away from the normal)
c) The effect of a prism has little practical use, for an optician, beyond theoretical discussions
d) Prisms are the fundamental building blocks of all eyeglass lenses

[#071] In the diagram, which letter represents the <u>real</u> position of the cloud relative to the observer?

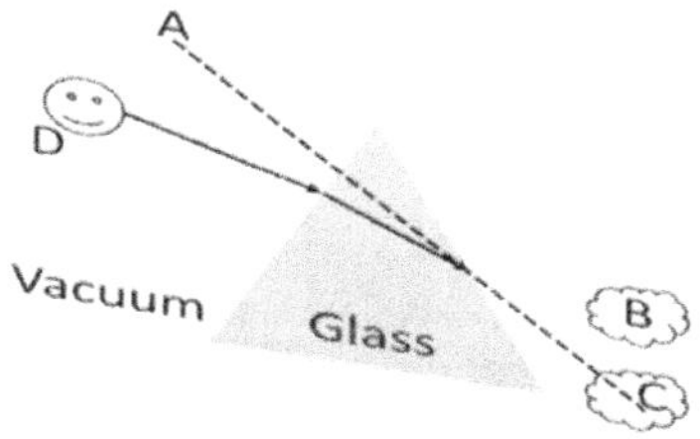

a) A
b) B
c) C
d) D

[#072] In the diagram, which letter represents the position of the cloud resulting from the effect of the prism?

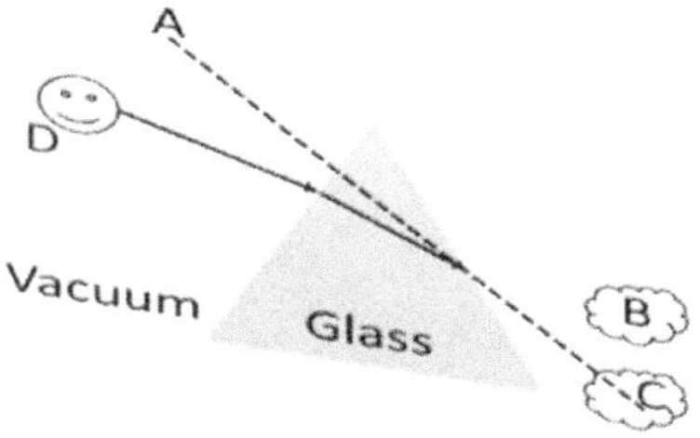

a) A
b) B
c) C
d) D

[#073] If you were observing the cloud from the letter D position, which letter represents the actual position of the cloud from the observer's position at D?

a) B
b) C
c) D
d) A

[#074] Which statement is **<u>NOT</u>** correct?

a) The apparent displacement of the cloud image occurs because the prism bends the light towards the base.
b) The information, from statements made in answers a and c of this question, are a theoretical optics exercise that has little relevance to making eyeglasses
c) When the prism bends the light towards the base the image "moves" or appears displaced towards the apex of the prism
d) The information, from statements made in answers a and c, is a fundamental optical building block in helping patients see better

[#075] Knowing the effect of a prism is only useful in rare cases, i.e., when the patient has a turned eye or related pathology.

a) True
b) False

Chapter Five Practice Exam

[#076] When you put two prisms together, either base-to-base or apex-to-apex you get:

a) A periscope prism

b) A telescope

c) A lens

d) An occluder

[#077] When you place two prisms together, base-to-base, then incident light rays shall:

a) Remain parallel

b) Diverge

c) Converge

d) Reflect

[#078] When you place two prisms together, apex-to-apex, you get:

a) A minus power lens

b) A plano lens

c) A progressive lens

d) A plus powered lens

[#079] When you place two prisms together, base-to-base, you get:

a) A minus power lens

b) A plano lens

c) A progressive lens

d) A plus power lens

[#080] When you place two prisms together, apex-to-apex, then incident light rays shall:

a) Remain parallel

b) Diverge

c) Converge

d) Reflect

[#081] When you place two prisms together, apex-to-apex, the result will be a profile that will:

a) Be thinner on the edges when compared to the center point

b) Be able to maintain consistent thickness from edge to center

c) Create parallel light waves

d) Be thicker on the edges when compared to the center point

[#082] When you place two prisms together, base-to-base, the result will be a profile that will be:

a) Able to create a pinhole effect

b) Able to diverge light rays

c) Able to converge light rays

d) Able to negate a pinhole effect

[#083] When you place two prisms together, apex-to-apex, the result will be a profile that will be:

a) Able to create a pinhole effect

b) Able to diverge light rays

c) Able to converge light rays

d) Able to form a real image

[#084] When you place two prisms together, base-to-base, the result will be a profile that will be:

a) Be thinner on the edges when compared to the center point

b) Be able to maintain consistent thickness from edge to center

c) Able to produce a virtual image

d) Be thicker on the edges when compared to the center point

[#085] Which statement is NOT correct?

a) Base-to-base prisms converge light and form a real image

b) Apex-to-apex prisms diverge light and form a virtual image

c) Apex-to-apex prisms are thinner on the edges and base-to-base prisms are thicker on the edges

d) Apex-to-apex prisms are thicker on the edges and base-to-base prisms are thinner on the edges

[#086] Which statement is NOT correct?

a) When you follow the path of light waves through base-to-base prisms, the waves diverge and create a virtual focus point

b) When you follow the path of light waves through base-to-base prisms, the waves converge and create a real focus point

c) When you follow the path of light waves through apex-to-apex prisms, they diverge and create a virtual focus point

d) Plus lenses converge light and minus lenses diverge light

[#087] The spot where the path of diverging light rays (from a lens) can be traced back to the point where the rays meet is called the:

a) Hypothetical focus point

b) Real focus point

c) Reflected focus point

d) Virtual focus point

[#088] Which statement is NOT correct?

a) The optical center (OC) is the point where two prisms are put together, both apex-to-apex or base-to-base

b) Regardless of the power and curvature of the two prisms, light shall neither converge or diverge at the OC

c) In a pair of eyeglasses, the position of the OC is of minor consideration if the Rx power is OK

d) Light rays shall pass through the OC without deviation

[#089] Which statement is NOT correct?

a) The joining point of two prisms, either apex-to-apex or base-to-base, creates an OC

b) Base-to-base prisms form a plus lens and the OC shall be at the thickest point

c) Apex-to-apex prisms form a minus lens and the OC shall be at the thinnest point

d) The relationship between a patient's eyes and the positioning of the OC is not typically critical if the Rx is correct

Chapter Six Practice Exam

[#090] The diagram is a representation of:

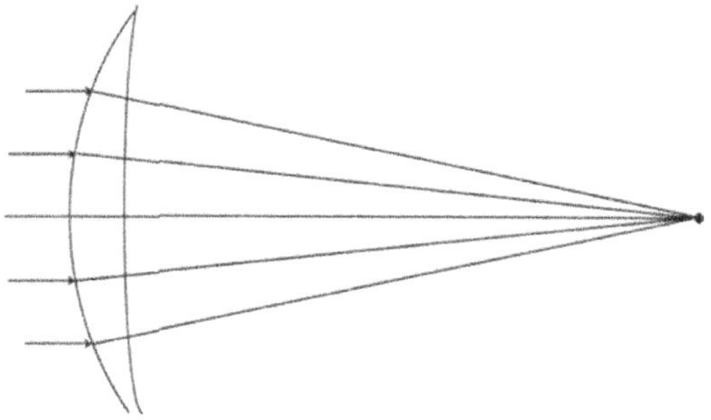

a) A minus lens diverging light
b) A convex lens diverging light
c) A minus lens converging light
d) A convex lens converging light

[#091] The distance from the lens to the focus point is called:

a) Focus length
b) A diopter
c) The OC distance
d) Focal length

[#092] When calculating the power of a lens the distance from the lens to the focus point should be expressed in the metric unit of:

a) Nanometers
b) Millimeters
c) Centimeters
d) Meters

[#093] If the distance from a convex lens to its focal point is one meter, the diopter power is:

a) -1.00D
b) -5.00D
c) +5.00D
d) +1.00D

$D = 1/d$

D = Diopter Power

D = distance in meters

[#094] If the distance from a convex lens to its focal point is 0.2 meters, the diopter power is:

a) -1.00D
b) -5.00D
c) +5.00D
d) +1.00D

[#095] If the distance from a concave lens to its virtual focal point is 0.2 meters, the diopter power is:

a) -1.00D
b) -5.00D
c) +5.00D
d) +1.00D

[#096] If the distance from a concave lens to its virtual focal point is one meter, the diopter power is:

a) -1.00D
b) -5.00D
c) +5.00D
d) +1.00D

[#097] If a plus lens has the focal length of 2.0 meters, the diopter power is:

a) -0.50D
b) -2.00D
c) +2.00D
d) +0.50D

[#098] If a plus lens has the focal length of 0.5 meters, the diopter power is:

a) -0.50D
b) -2.00D
c) +2.00D
d) +0.50D

[#099] If a convex lens has the focal length of 0.25 meters, the diopter power is:

a) +4.00D
b) -4.00D
c) +0.50D
d) -0.50D

[#100] If a plus lens has the focal length of 2.00 meters, the diopter power is:

a) +4.00D
b) -4.00D
c) +0.50D
d) -0.50D

[#101] If a minus lens has a focal length of one-tenth of a meter, the diopter power is:

a) -1.00D
b) +1.00D
c) -10.00D
d) +10.00D

[#102] Which statement is **<u>NOT</u>** correct?

a) A minus lens does not bring light to a point focus as a plus lens does. The purpose of a minus lens is to refract light rays in a diverging fashion
b) To calculate the diopter power of a minus lens we must use a virtual or negative focal length
c) The math for a negative focal length demands we use a negative distance (-d) measurement in the formula
d) As the focal length increases so does the diopter power

[#103] Which statement is **<u>NOT</u>** correct?

a) Real images form where light converges.
b) Virtual images form (virtually) where light appears to materialize.
c) You may compute the focal length to the virtual image with the formula D = 1/-d.
d) The stronger the diopter power, the longer the focal length.

[#104] If the front curvature of a lens is +10.00D and the back surface curvature is -2.00D, what is the nominal lens power?

a) +8.00D
b) +7.00D
c) +6.00D
d) +5.00D

[#105] If the front curvature of a lens is +12.00D and the back surface curvature is -4.00D, what is the nominal lens power?

a) +8.00D
b) +7.00D
c) +6.00D
d) +5.00D

[#106] If the front curvature of a lens is +6.00D and the back surface curvature is -1.00D, what is the nominal lens power?

a) +8.00D
b) +7.00D
c) +6.00D
d) +5.00D

[#107] If the front curvature of a lens is +8.00D and the back surface curvature is -2.00D, what is the nominal lens power?

a) +8.00D
b) +7.00D
c) +6.00D
d) +5.00D

[#108] With the use of a lens clock the optician can determine:

a) The exact power of a lens
b) The base curve of a patient's old lenses
c) The type of PAL the patient is wearing
d) The minimum blank size needed

[#109] Vertex distance is:

a) The distance between your patient's eyes
b) The diameter of a lens blank
c) The longest focal length of a pair of lenses
d) The distance between the front of the patient's eye (cornea) and the back surface of a lens (Effect must often be compensated for when contact lenses are involved or eyeglasses typically in high powers and unusual fitting distance.)

[#110] Which statement is **NOT** correct?

a) Moving a plus lens closer to your eye weakens the effective resultant power
b) Moving a plus lens farther away from your eye makes the effective resultant power stronger
c) Increasing the vertex distance reduces power to a plus lens and adds minus power to a minus lens
d) Increasing the vertex distance adds power to a plus lens and weakens a minus lens

[#111] The doctor prescribes a +11.50D lens. The refraction was done at 12 mm. What is the effective power at 6 mm?

a) +12.25D
b) +11.50D
c) +10.75D
d) +10.00D

Rx = Prescribed Power
d = distance change in meters (+/-)
Rrx = Resultant Power
Rx / 1 + d x Rx = Rrx

[#112] The doctor prescribes a -11.50D lens. The refraction was done at 12 mm. What is the effective power at 6 mm?

a) -12.35D
b) -11.50D
c) -10.75D
d) -10.00D

[#113] The doctor prescribes a -9.25D lens. The refraction was done at 10 mm. What is the effective power at 14 mm?

a) -12.35D
b) -9.25D
c) -8.92D
d) -7.62D

[#114] The doctor prescribes a -7.00D lens. The refraction was done at 8 mm. What is the effective power at 14 mm?

a) -12.35D
b) -9.25D
c) -8.92D
d) -6.73D

[#115] The nominal power of a lens is -5.50D. The front surface power, as per the lens clock, is +2.00. What is the back surface power?

a) -7.50D
b) +7.50D
c) -5.50D
d) +5.50D

[#116] When calculating the resultant power from a change in vertex distance:

a) You must use a negative number for distance (d) when the vertex distance increases and a positive number when vertex distance is reduced
b) You must use a positive number for distance (d) when the vertex distance increases and a negative number when vertex distance is reduced
c) You must use a positive number for distance (d) when the vertex distance increases and a positive number when vertex distance is reduced
d) You must use a negative number for distance

[#117] A post-cataract patient with failed intraocular implants has been prescribed +18.00D in both eyes (OU). The doctor refracted the patient at 12 mm and asked you to compensate the power based on the vertex distance you measure with a distometer. Your measurement establishes a vertex distance of 8 mm with the frames the patient selected. What power will you order from the lab? Round to the nearest eighth of a diopter (.12).

Rx = Prescribed Power
D = distance change in Meters
Rrx = Resultant Power
Rx/1 + d x Rx = Rrx

a) +18.00D
b) -16.25D
c) +16.75D
d) +19.37D

[#118] A -7.75D lens has been fitted at a distance change of -.004 (negative, away). The resultant power is:

a) +7.52
b) -7.87
c) +7.87
d) -7.52

[#119] Which statement is **NOT** correct?

a) The focal length of a +1.00D lens is 1.0 m
b) The focal length of a +2.00D lens is 0.50 m
c) The focal length of a +4.00D lens is 0.250 m
d) The focal length of a +8.00D lens is 0.120 m

[#120] Which statement is **NOT** correct?

a) A concave lens that has an increasing or longer vertex distance becomes a weaker resultant power
b) A convex lens that has an increasing or longer vertex distance becomes a stronger resultant power
c) When troubleshooting a patient complaint (i.e. distortion, blurriness, headaches with new glasses, etc.), determining the front curve of the old lenses with a lens clock is a prudent step
d) When a lens gets stronger its focal length increases in a linear fashion with the power

Chapter Seven Practice Exam

[#121] Which statement is NOT correct?

a) A spherical lens has the same shape characteristics as if a slice off of a round ball

b) A toric lens has the same shape characteristics as if you took a slice off of a football

c) A spherical lens may be rotated clockwise or counter-clockwise with no apparent change in power or image movement

d) A toric lens has the same power throughout the lens

[#122] The optical cross is used to provide the optical industry with a uniform system to visualize a patient's Rx relative to the positioning of eyeglasses.

For the patient Rx of +2.5o diopters spherical right eye (OD), specify the power of the lens in the vertical meridian?

a) +2.50 @ 90o

b) +2.50 @ 360

c) +2.50 @ 180o

d) +1.25 @ 90o

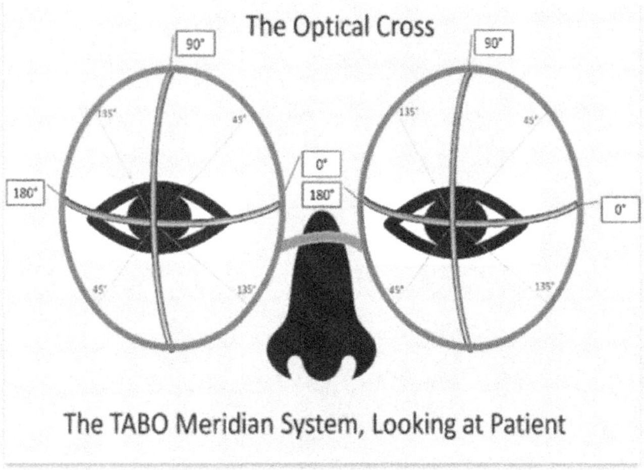

[#123] Which statement is NOT correct?

a) In a spherical lens the power will be identical from zero to 180o

b) 180o is horizontal

c) 45o is oblique

d) In a toric lens the power will be identical from zero to 90o

[#124] Which statement is NOT correct

a) The power in a toric lens changes incrementally at every degree on the optical cross from zero to 180

b) The first number in a Rx is the base power and is called "sphere"

c) A toric lens typically maintains a consistent edge thickness at the circumference of the lens

d) The second number in the Rx is added (algebraically) to the sphere (sph) and is the full cylinder (cyl) power

[#125] Which statement is NOT correct?

a) If you take a toric lens blank and look through it while rotating it, repeatedly, clockwise, and counter-clockwise, you will see a distinct "scissors-like" motion; particularly on stronger cylinder powers

b) With a toric surface like the side of a barrel, one curvature is the long axis and a steeper curvature is the shorter axis

c) The best practice is to use the spherical equivalent when the power is low

d) A toric lens will have differing edge thicknesses. The edge thickness difference on a minus toric lens is particularly noticeable with high cylinder powers

[#126] What is the power of the OD toric lens at 90 degrees (vertical)?

	Sphere	Cylinder	Axis
OD – Right	+1.00	-1.00	180
OS - Left	+1.00	-1.00	180

a) Plano

b) +1.00

c) -1.00

d) 180

[#127] What is the power of the OD toric lens at 90° and the OS power at 180°?

	Sphere	Cylinder	Axis
OD – Right	+1.00	-2.50	180
OS - Left	+1.00	-1.00	180

a) OD Plano @ 90, OS Plano @ 180

b) OD +1.00 @ 90, OS -1.00 @ 180

c) OD -1.50 @ 90, OS +1.00 @ 180

d) OD -1.50 @ 90, OS Plano @ 180

[#128] What is the power of the OD toric lens at 90° and the OS power at 180°?

a) OD -3.50 @ 90, OS -1.50 @ 180

b) OD -2.50 @ 90, OS -3.00 @ 180

c) OD +3.50 @ 90, OS -1.50 @ 180

d) OD -3.50 @ 90, OS Plano @ 180

	Sphere	Cylinder	Axis
OD - Right	-1.00	-2.50	180
OS - Left	-1.50	-3.00	180

[#129] What is the power of the OD toric lens at 90° and the OS power at 180°?

a) OD -1.00 @ 90, OS -1.50 @ 180

b) OD -2.50 @ 90, OS -3.00 @ 180

c) OD +3.50 @ 90, OS -1.50 @ 180

d) OD –1.00 @ 90, OS -4.50 @ 180

	Sphere	Cylinder	Axis
OD – Right	-1.00	-2.50	90
OS - Left	-1.50	-3.00	90

[#130] What is the power of the OD toric lens at 90° and the OS power at 180°?

a) OD +0.50 @ 90, OS Plano @ 180

b) OD +0.50 @ 90, OS -3.00 @ 180

c) OD +4.25 @ 90, OS -3.00 @ 180

d) OD +1.75 @ 90, OS -4.50 @ 180

	Sphere	Cylinder	Axis
OD – Right	+1.75	-2.50	45
OS - Left	+1.50	-3.00	45

[#131] What is the power of the OD toric lens at 90° and the OS power at 180°?

a) OD +0.50 @ 90, OS Plano @ 180

b) OD +0.50 @ 90, OS -3.00 @ 180

c) OD +4.25 @ 90, OS -3.00 @ 180

d) OD +1.75 @ 90, OS -4.50 @ 180

	Sphere	Cylinder	Axis
OD – Right	+1.75	-2.50	135
OS - Left	+1.50	-3.00	135

This diagram illustrates the edge profile of lenses with various prescriptions.

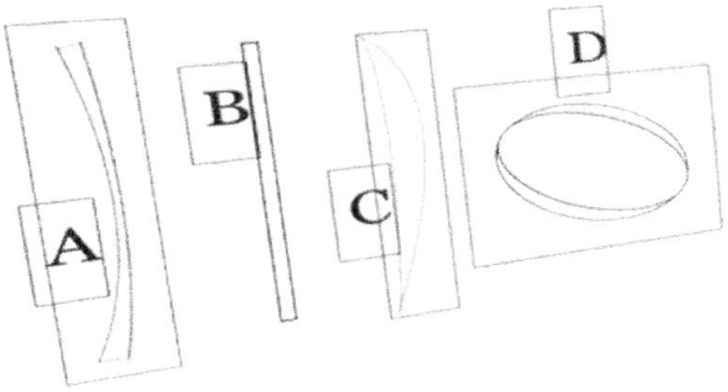

[#132] Which profile represents a Plano lens?

a) A
b) B
c) C
d) D

[#133] Which profile represents a minus lens?

a) A
b) B
c) C
d) D

[#134] Which profile represents a toric lens?

a) A
b) B
c) C
d) D

[#135] Which profile represents a plus lens?

a) A
b) B
c) C
d) D

[#136] Which statement accurately describes the difference between these two Rx's?

a) The plus cylinder Rx is stronger than the minus cylinder Rx

b) The plus cylinder Rx is weaker than the minus cylinder Rx

c) The differing axis location makes the powers different

d) The two Rx's are exactly the same in power but only differ in format

	SPH	CYL	AXIS
R	-5.00	-2.00	180
L	-5.00	-2.00	180

	SPH	CYL	AXIS
R	-7.00	+2.00	90
R	-7.00	+2.00	90

[#137] Which statement is NOT correct?

a) The sphere power is as written at the axis specified

b) 90 degrees from the sphere power is the sphere power added to the cylinder power

c) On a toric lens the power is different on every degree from zero to 180

d) The power of the lenses in the minus cylinder format is different than the power of the lenses in the plus cylinder format in the preceding examples

Chapter Eight Practice Test

[#138] The human eye has many similarities to a:

a) Computer
b) Camera
c) Calculator
d) Cytoplasm

[#139] The first lens of the human eye is the:

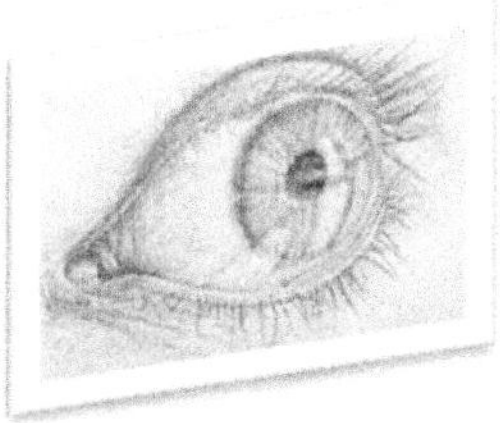

a) Iris
b) Visual axis
c) Cornea
d) Pupil

[#140] The primary optical purpose of the cornea is to provide:

a) 40 to 45 diopters of light diverging power
b) 40 to 45 diopters of light converging power
c) 35 to 50 diopters of light diverging power
d) 35 to 50 diopters of light converging power

[#141] The exterior surface of the cornea is constantly bathed in an oily tear layer produced by the lacrimal gland and drained away by the tear duct. What other fluids are critical to the function of the eye?

a) Zonules and cortex
b) Fovea and choroid
c) Aqueous and vitreous humor
d) Choroid and spinal fluid

[#142] Select the answer that correctly sequences the anatomical parts in order from front to back:

a) Optic nerve, vitreous, crystalline lens, pupil, anterior chamber, cornea
b) Optical axis, ciliary body, optical axis, Zonules
c) Cornea, anterior chamber, pupil, crystalline lens, vitreous, optic nerve
d) Cornea, cortex, pupil, iris, fovea, sclera

[#143] Select the letter that points to the retina:

a) E
b) F
c) D
d) A

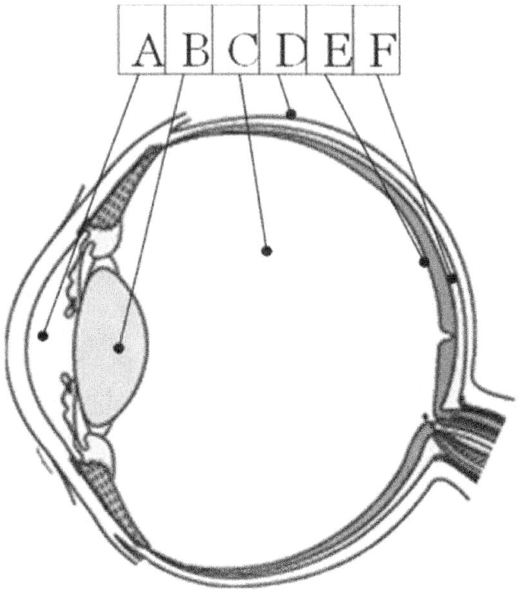

[#144] Select the letter that points to the vitreous humor:

a) E
b) D
c) C
d) A

[#145] Select the letter that points to the crystalline lens:

a) B
b) C
c) D
d) E

[#146] Select the letter that points to the anterior chamber:

a) A
b) B
c) C
d) D

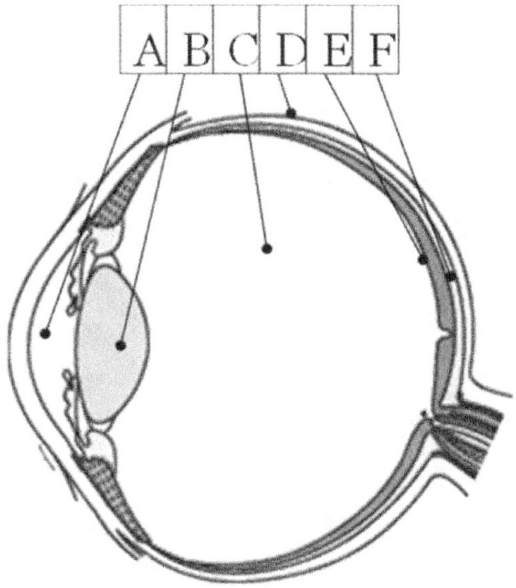

[#147] Select the letter that points to the choroid

a) C
b) D
c) F
d) E

[#148] Select the letter that points to the sclera:

a) C
b) D
c) E
d) F

[#149] Which statement is **NOT** correct?

a) The optical axis is a line that illustrates the geometric central axis through the eye. It is sometimes called the principal axis
b) The visual axis is a line from the center of the crystalline lens to the densest area of photoreceptor concentration
c) The sclera is a white, tough fibrous outer layer that protects the eye
d) The fovea is the point where the optic nerve exits the eyeball and is the greatest cluster of photoreceptors in the eye

[#150] Select the letter that points to the optic nerve:

a) D
b) A
c) F
d) E

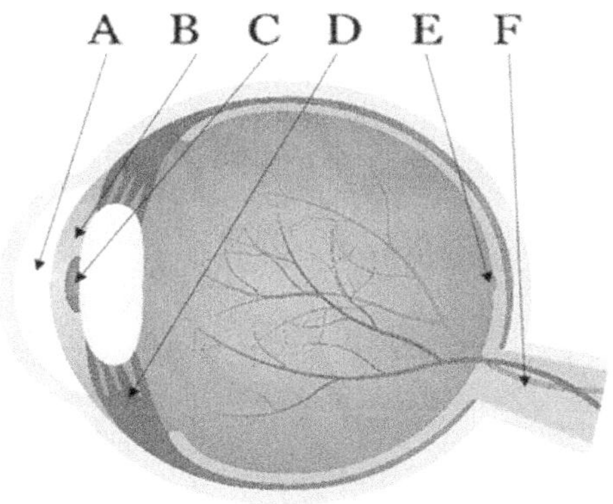

[#151] Select the letter that points to the fovea:

a) D
b) A
c) F
d) E

[#152] Select the letter that points to the aqueous humor:

a) A
b) B
c) D
d) E

[#153] Select the letter that points to the iris:

a) A
b) B
c) C
d) D

[#154] Select the letter that points to the zonular fibers:

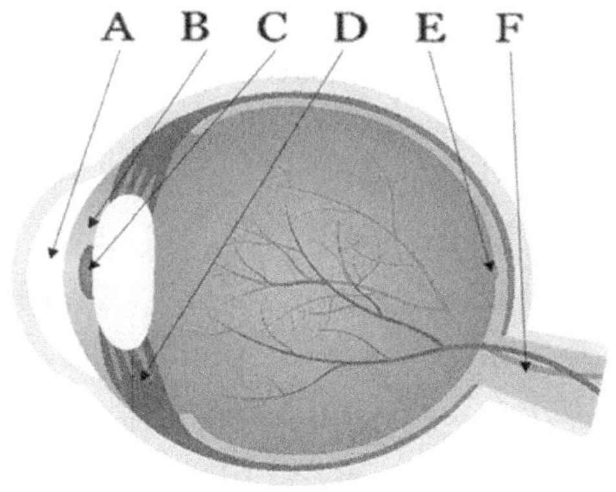

a) A
b) B
c) C
d) D

[#155] Select the letter that points to the pupil:

a) A
b) B
c) C
d) D

[#155] Select the letter that points to the pupil

e) A
f) B
g) C
h) D

[#156] Which statement is **NOT** correct?

a) The cornea refracts light, with 40 to 45 diopters of plus power, into the crystalline lens by way of the aqueous humor
b) The iris reacts to lighting levels to make the pupil larger or smaller as needed
c) Light rays travel along the optical axis to reach the fovea
d) The pupil serves to create a pinhole effect allowing parallel light rays, as opposed to scattered light rays, to land on the retina

[#157] Which statement is **NOT** correct?

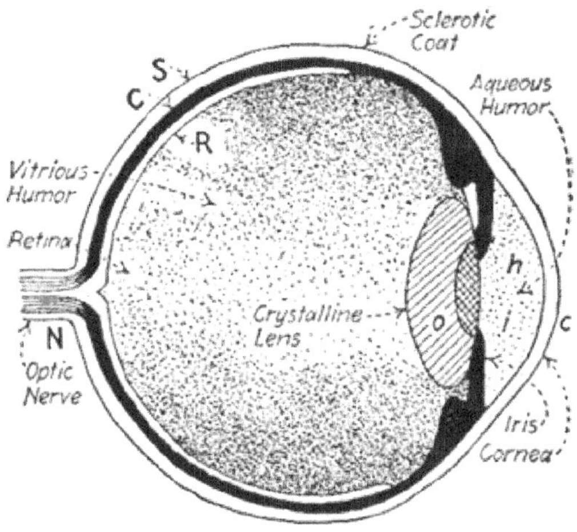

a) The retinal cones predominate at the edges of the retina while the rods are concentrated near the center
b) The photoreceptor cells (rods & cones) are like the electro-photo sensors in your cell phone camera
c) The choroid provides nourishment to the retina with a layer of blood vessels
d) The retinal rods predominate the edges of the retina while the cones are more concentrated near the center

[#158} Which statement is **NOT** correct?

a) The ring of ligaments (zonular fibers in tandem with ciliary body), surrounding the edges of the crystalline lens tighten or relax as needed for near focusing
b) The crystalline lens is typically 16 to 20 diopters
c) The action of focusing for near vision is called accommodation.
d) Focusing ability is the primary function of the biconcave or minus powered crystalline lens

[#159] Which statement is **NOT** correct?

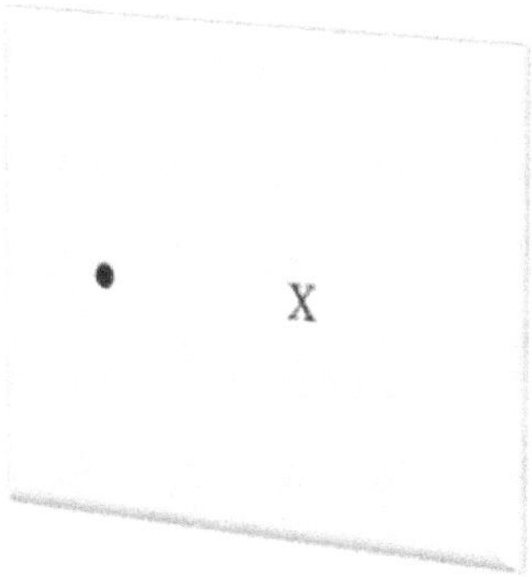

a) The blind spot on the retina reveals where the optic nerve exits the eye
b) There are vision deficiencies, such as the blind spot that the visual cortex in the brain, corrects for us
c) The blind spot can get worse over time
d) The optic nerves from each eye split and the fibers cross each other at the optical chiasm

[#160] What letter is the inferior rectus?

a) B
b) C
c) D
d) E

[#161] What letter is the superior rectus?

a) A
b) B
c) D
d) F

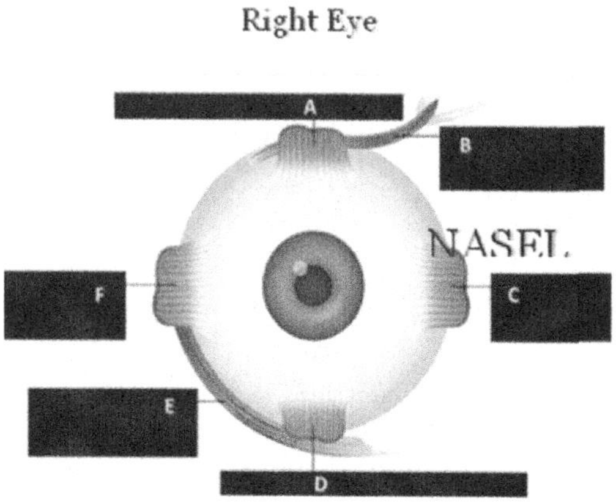

[#162] What letter is the inferior oblique?

a) F
b) E
c) D
d) B

[#163] What letter is the medial rectus?

a) B
b) C
c) D
d) E

[#164] What letter is the superior oblique?

a) F
b) C
c) B
d) E

[#165] What letter is the lateral rectus?

a) C
b) D
c) E
d) F

[#166] What letter is the frontal bone

a) E
b) A
c) B
d) D

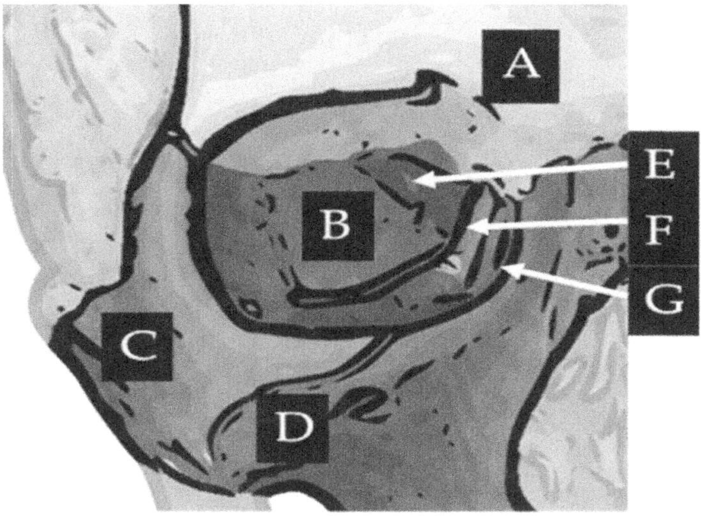

[#167} What letter is the palatine bone?

a) F
b) E
c) D
d) C

[#168] What letter is the sphenoid bone?

a) D
b) C
c) B
d) A

[#169] What letter is the lacrimal bone?

a) G
b) E
c) C
d) A

[#170] What letter is the maxillary bone?

a) D
b) E
c) F
d) A

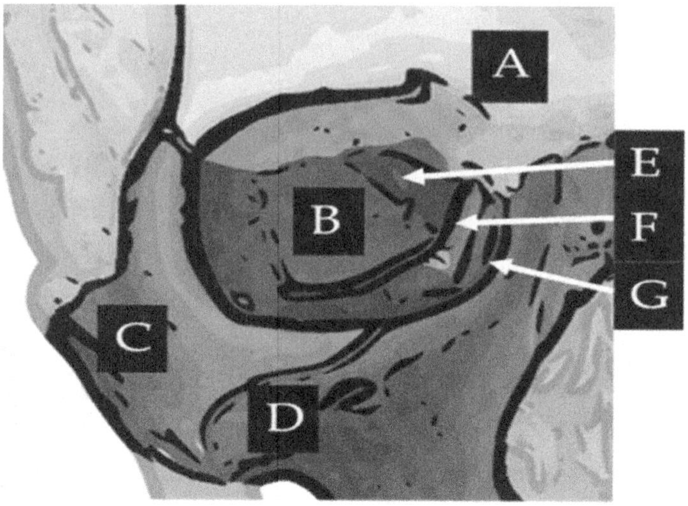

[#171] What letter is the ethmoid bone?

a) E
b) D
c) B
d) A

[#172] What letter is the zygomatic bone?

a) B
b) C
c) D
d) E

[#173] What eye movement does the inferior oblique control?

a) Upward and outward
b) Downward and outward
c) Downward
d) Outward

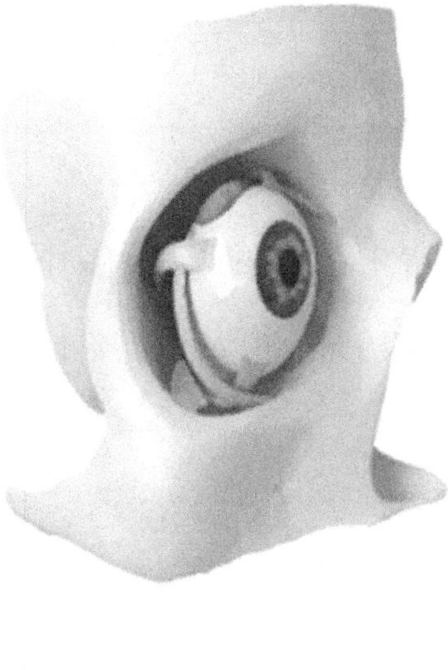

[#174] What eye movement does the inferior rectus control?

a) Upward and outward
b) Downward and outward
c) Downward
d) Outward

[#175] What eye movement does the superior rectus control?

a) Upward and outward
b) Upward
c) Downward
d) Outward

[#176] What eye movement does the superior oblique control?

a) Upward and outward
b) Upward
c) Downward and outward
d) Outward

[#177] What eye movement does the lateral rectus control?

a) Upward and outward
b) Inward
c) Downward and outward
d) Outward

[#178] What eye movement does the medial rectus control?

a) Upward and outward
b) Inward
c) Downward and outward
d) Outward

Chapter Nine Practice Test

Myopia	Hyperopia	Astigmatism	Presbyopia	Aphakia or Post-Cataract
		Mixed		
		Simple		
		Compound		
		Oblique		

[#179] Which statement is **NOT** correct?

a) Near-sighted people can see up close

b) Myopia occurs when light rays converge to a point focus prior to reaching the retina

c) Myopia occurs when light rays converge to a point focus after reaching the retina

d) Myopia typically needs a minus lens to correct the refractive error of the eye

[#180] Which statement is **NOT** correct?

a) Far-sighted people typically can see far away

b) Hyperopia occurs when light rays converge to a point focus prior to reaching the retina

c) Hyperopia occurs when light rays attempt to converge to a point focus after reaching the retina

d) Hyperopia typically needs a plus lens to correct the error of the eye

[#181] Which statement is **NOT** correct?

a) Mixed astigmatism results when a plus cylinder Rx is used

b) With mixed astigmatism the error of the eye results in focal points that are both before and after the retina

c) Mixed astigmatism typically needs a toric lens to correct the refractive error of the eye

d) Astigmatism causes substandard uncorrected acuity at all distances

[#182] Which statement is **NOT** correct?

a) Simple myopic astigmatism results in one focus point on the retina and another focus point before the retina

b) Simple myopic astigmatism requires a toric lens to correct

c) Astigmatic refractive error is typically created either in the cornea or the crystalline lens

d) A lens with a spherical equivalent Rx will fully correct simple myopic astigmatism

[#183] Which statement is **NOT** correct?

a) Simple hyperopic astigmatism typically needs a concave form of lens to correct

b) Simple hyperopic astigmatism results in one focus point on the retina and another focus point after or behind the retina

c) Simple hyperopic astigmatism typically needs a toric form of lens to correct

d) Near vision is typically compromised with uncorrected simple hyperopic astigmatism

[#184] Which statement is **<u>NOT</u>** correct?

a) Compound hyperopic astigmatism typically is the result of both primary focal points attempting to manifest behind the retina

b) Near vision is typically compromised with uncorrected compound hyperopic astigmatism

c) Compound hyperopic astigmatism can be corrected with a convex toric form of lens

d) Simple over-the-counter reading glasses will optimally correct compound hyperopic astigmatism

[#185] Which statement is **<u>NOT</u>** correct?

a) Compound myopic astigmatism is the result of both primary focal points manifesting before the retina

b) Distance vision is compromised with compound myopic astigmatism

c) Compound myopic astigmatism can be corrected with a convex toric form of lens

d) Compound myopic astigmatism can be corrected with a concave toric form of lens

[#186] What type of astigmatism is represented by this Rx?

SPH	CYL	AXIS
+2.25	-5.25	180

a) Against-the-Rule astigmatism

b) With-the-Rule astigmatism

c) Oblique astigmatism

d) Irregular astigmatism

[#187] What type of astigmatism is represented by this Rx?

SPH	CYL	AXIS
+0.25	-4.00	20

a) Against-the-Rule astigmatism

b) With-the-Rule astigmatism

c) Oblique astigmatism

d) Irregular astigmatism

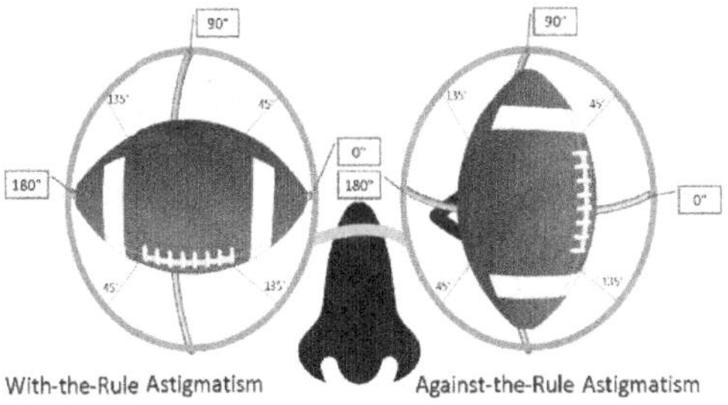

[#188] What type of astigmatism is represented by this Rx?

SPH	CYL	AXIS
-1.25	-4.00	80

a) Against-the-Rule astigmatism

b) With-the-Rule astigmatism

c) Oblique astigmatism

d) Irregular astigmatism

[#189] What type of astigmatism is represented by this Rx?

SPH	CYL	AXIS
+5.25	-2.00	60

a) Against-the-Rule astigmatism

b) With-the-Rule astigmatism

c) Oblique astigmatism

d) Irregular astigmatism

[#190] What type of astigmatism is represented by this Rx?

SPH	CYL	AXIS
+5.25	-2.00	59

a) Against-the-Rule astigmatism

b) With-the-Rule astigmatism

c) Oblique astigmatism

d) Irregular astigmatism

[#191] What type of astigmatism is represented by this Rx?

SPH	CYL	AXIS
-3.00 (sph power- OD)		32
	-5.00 (cyl power - OD)	63

a) Against-the-Rule astigmatism

b) With-the-Rule astigmatism

c) Oblique astigmatism

d) Irregular astigmatism

[#192] What type of astigmatism is represented by this Rx?

SPH	CYL	AXIS
-13.00	-1.00	29

a) Against-the-Rule astigmatism

b) With-the-Rule astigmatism

c) Oblique astigmatism

d) Irregular astigmatism

[#193] Which statement is **NOT** correct?

a) Presbyopia is the loss of accommodation (declining near vision)

b) Presbyopia typically occurs after the age of 40

c) There are a number of corrective combinations for presbyopia (readers, bifocal, etc.)

d) Presbyopia and hyperopia are the same thing

[#194] Due to a lessening of elasticity over time this portion of the eye creates the condition of presbyopia.

a) Cornea

b) Pupil

c) Crystalline lens

d) Iris

[#195] What is a "PAL"?

a) Trifocal

b) Flat top bifocal

c) Single vision

d) Progressive addition lens

[#196] Which statement is **NOT** correct?

a) A cataract occurs in the crystalline lens and, over time, the lens becomes cloudy/opaque

b) A cataract that has progressed to the point of obstructed vision will cause decreased vision, halos, and loss of color perception

c) Typically, the cloudy lens is removed surgically and glasses will be needed to replace the diopter power of the removed crystalline lens

d) Cataracts can be caused, on rare occasions, by injury

Chapter Ten Practice Test

[#197] The index of refraction (n-value) of a vacuum is:

a) 1.49

b) 1.523

c) 1.33

d) 1.00

n = c / v
• n = Index of Refraction
• c = speed of light in a vacuum (299,792,458 meters/sec) OR
• c = speed of light in a vacuum (186,000 miles/sec
• v = speed of light in a transparent material

[#198] Light slows down in a clear material like glass. How is the n-value affected on different clear materials?

a) Denser materials that slow light more will have a higher n-value

b) Denser materials that slow light more will have a lower n-value

c) Different densities do not affect n-value

d) Less dense materials reflect light

[#199] A trivex lens has an n-value of 1.53. What is the speed of light in this material?

a) 195,942,783 m/s; 121,569 miles/s

b) 194,842.783 m/s; 124,767 miles/s

c) 121,567 m/s; 195,943 miles/s

d) None of the above

[#200] Which statement is <u>**NOT**</u> correct?

a) The denser the transparent material, the more light is slowed down

b) The more that light slows down in a lens the greater the light will refract

c) Denser lens materials allow flatter curves and a thinner lens

d) The amount of refraction and dispersion is independent of the n-value

[#201] If light travels at a speed of 117,276.12 miles per second in a lens material, which lens material is it?

a) Polycarbonate

b) Trivex

c) Flint Glass

d) CR-39

Typical n values

- Vacuum = 1.00
- Water = 1.33
- Plastic Lenses (CR-39) = 1.498
- Spectacle Crown Glass = 1.523
- Trivex = 1.53
- Polycarbonate = 1.586
- Flint Glass = 1.61
- Hi-Index Plastic = 1.67 or 1.74

[#202] Which statement is **NOT** correct?

a) Abbe value assigns a numerical value to describe the amount of dispersion inherent in a lens material

b) CR-39 and crown glass have less dispersion than polycarbonate or hi-index plastics

c) The higher the Abbe value the more dispersion occurs

d) The higher the Abbe value the less dispersion occurs

[#203] Abbe value should be considered when the patient:

a) Has a strong prescription and wants large eyeglasses

b) Is selecting sunglasses

c) Needs SV readers

d) Has Against-the-Rule astigmatism

[#204] This lens material with an Abbe value of 59 that was widely used into the 1980s, has good optics and good scratch resistance, but was often too heavy to be comfortable is:

a) Hi-lite glass

b) Trivex

c) Crown glass

d) CR-39

[#205] The lens material with an Abbe value of 30 has exceptional impact resistance and inherently blocks UV without additional treatments is:

a) Polycarbonate
b) CR-39
c) Crown glass
d) Trivex

[#206] The lens material with an Abbe value of 32 to 36 (material has several n-values of 1.67 to 1.74) is noticeably thinner in higher prescriptions but tends to be more expensive is:

a) CR-39
b) Hi-lite glass
c) Hi-index plastic
d) Trivex

[#207] The lens material that is considered a premium polycarbonate product with better optics, has an excellent Abbe value of 45 with superior impact resistance and inherently blocks UV is:

a) CR-39
b) Hi-lite glass
c) Hi-index plastic
d) Trivex

[#208] The lens material with an Abbe value of 58 that has good optics (better than poly) but is thicker than poly and scratches easily without a hard coat is:

a) CR-39
b) Hi-lite glass
c) Hi-index plastic
d) Trivex

Chapter Eleven Practice Test

[#209] Which statement is **NOT** correct?

a) Single Vision lenses (SV) come in a variety of profiles including spherical, aspherical, myodisc, and lenticular

b) SV lenses are designed for specific use. Distance correction for myopic patients is the most common, but they can also be task-specific as in computer distance or near only

c) The myodisc and lenticular forms are reserved for very strong Rxs typically above 15.00 diopters

d) A pair of SV eyeglasses may be used to correct all errors inherent to a myopic presbyopic patient (absent look-unders)

[#210] Which statement is **NOT** correct?

a) All SV lenses are spherical and deliver a specific correction for one specific distance (far, intermediate, near etc.)

b) Hybrid SV designs have been developed that allow more than one specific distance use

c) There are SV lenses that blend aspheric design with modest progressive power towards the edges in the positive diopter power direction

d) The SV hybrids help patients who have demonstrated accommodative deficiencies

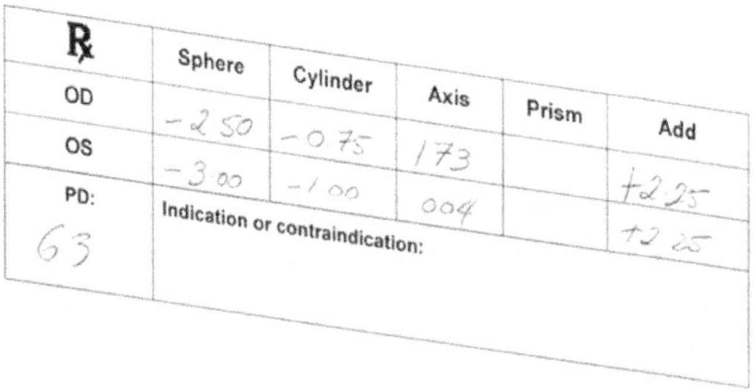

℞	Sphere	Cylinder	Axis	Prism	Add
OD	-2.50	-0.75	173		+2.25
OS	-3.00	-1.00	004		+2.25
PD: 63	Indication or contraindication:				

[#211] Which statement is **NOT** correct?

a) A prescription with an ADD power typically calls for some type of multifocal (or can be combined into SV for a specific task)

b) There is a wide range of multifocal options to choose from that can be designed for everyday use, a specific task or an occupational need

c) One drawback to a multifocal is that when you need to see close above eye level you are looking through the distance portion of the lens. No multifocal will help near vision above eye level

d) There are multifocals that have a visible line, a blended out line and "no-lines" that rely on gradually progressive power changes

[#212] The lens design with a line all the way across the lens (for a wider reading area) and that has dramatically decreased in popularity over the last twenty years is called:

a) A flattop

b) A fused bifocal

c) A draftsman's or executive lens

d) Freeform PAL

[#213] Which statement is **NOT** correct regarding the illustration with two D-shaped segments?

a) This is a little-used but necessary design for occupations that require hours of "above eye level" work

b) The lens is called the "Double-near only"

c) The lens is called the "Double-D"

d) Occupational lenses like this are designed for specific uses

[#214] Which statement is **NOT** correct?

a) Flattops come in a variety of widths such as 22 mm, 25 mm, 28 mm, 35 mm, 45 mm and the Exec
b) Round segs offer a more cosmetically appealing alternative to flattops
c) Blended bifocals are very similar to a PAL
d) The trifocal offers three distinct zones (distance, intermediate and near)

[#215] The distance portion of a PAL:

a) Is located roughly on the bottom third of the lens
b) Is located roughly on the middle third of the lens
c) Is located roughly on the top third of the lens
d) Is not part of the PAL design

[#216] The intermediate (arm's length) portion of a PAL:

a) Is located roughly on the bottom third of the lens
b) Is located roughly on the middle third of the lens
c) Is located roughly on the top third of the lens
d) Is not part of the PAL design

[#217] The near vision portion of a PAL:

a) Is located roughly on the bottom third of the lens
b) Is located roughly on the middle third of the lens
c) Is located roughly on the top third of the lens
d) Is not part of the PAL design

[#218] Which statement is **NOT** correct?

a) There is some inherent distortion to the sides of the central portion of a PAL but it has become an exceptionally useful tool to address presbyopia

b) Early hard PAL designs have passed the test of time and are rarely unacceptable to patients

c) Research and technology have smoothed the curve patterns that led to "soft" designs and now freeform lenses with an increase in patient acceptance

d) A PAL includes laser etchings that are critically important reference points to manufacture the glasses and maximize patient comfort

[#219] On this cut-out chart, what letter is the reference point for prism?

a) A
b) B
c) C
d) D
e) E
f) F

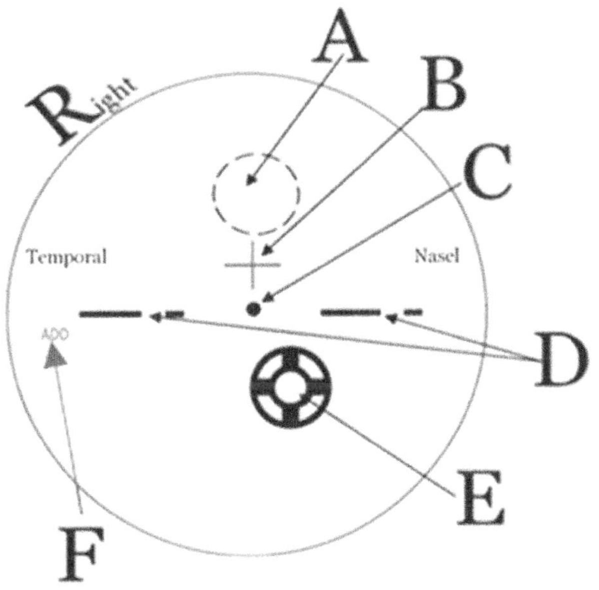

[#220] On this cut-out chart, what letter is the reference point for near vision?

a) A
b) B
c) C
d) D
e) E
f) F

[#221] On this cut-out chart, what letter is the reference point for distance vision?

a) A
b) B
c) C
d) D
e) E
f) F

[#222] On this cut-out chart, what letter is the reference point (laser etching) for the ADD?

a) A
b) B
c) C
d) D
e) E
f) F

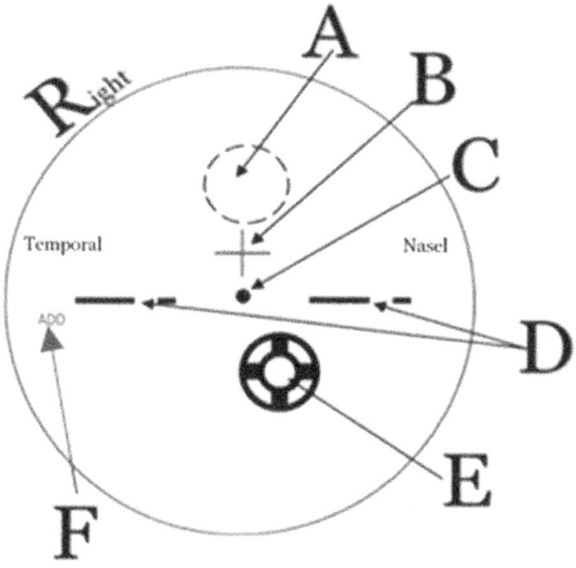

[#223] On this cut-out chart, what letter is the reference point for fitting the center pupil?

a) A
b) B
c) C
d) D
e) E
f) F

[#224] Which statement is **NOT** correct?

a) State-of-the-art PAL's are Freeform (digitally processed) lenses.
b) This computer-guided process creates both near and distance areas on the back surface of the freeform or digital lens and that bring the optics closer to the eye.
c) Near optics are wider and edge distortions are greatly reduced compared to older methods of production
d) Freeform availability is limited to "brand name" lenses

Chapter Twelve Practice Test

[#225] Which statement is **NOT** correct?

a) Lens with anti-reflective (ARC) properties greatly reduce the reflectivity inherent to all transparent mediums

b) The reduced reflectivity allows more light to be transmitted through the lens, a particular aid to night vision

c) A drawback to ARC is the reduced cosmetic appearance

d) Without ARC approximately 8% of light striking the lens is reflected

[#226] Which statement is **NOT** correct?

a) ARC does not change the apparent color of the lens

b) Some premium ARC products can build in enhanced blue light and/or UV protection

c) ARC must be applied to a clean lens, in a vacuum chamber and typically in multiple layers

d) ARC greatly improves the cosmetic appearance of the lens, particularly under harsh lighting

[#227] Blue light and UV are considered potentially harmful to your eyes and many products have been developed to absorb those wavelengths. What is the wavelength of the blue-violet light targeted by these products?

a) 415-455 nm

b) 445-485 nm

c) 720-760 nm

d) 735-775 nm

[#228] Mirror coatings are applied in a vacuum, like ARC. They come in a wide range of colors and are a popular feature, typically on sunglasses. Aside from cosmetics, what is the primary function of a mirror coat on glasses?

a) Cosmetics are the only function of ophthalmic mirrors

b) Mirrors increase the amount of light passing through the lens (transmission)

c) Mirrors decrease the amount, via reflection, of light passing through the lens

d) Mirrored suns can be turned around and can be used to make sure your hair is in order

[#229] What specific wavelengths activate a photosensitive lens?

a) Infrared waves
b) Yellow light waves
c) UV light waves
d) Total electromagnetic spectrum exposure is needed

[#230] Polarized sun lenses are useful in reducing reflected glare from damp surfaces such as a wet highway, water or ice. How does the polarization achieve this?

a) It is the lens material that slows light and blocks glare
b) As it is a tinted sun lens, harsh glare is blocked by the color
c) Polarization works by filtering scattered light into parallel beams, reducing glare
d) Polarization works in a similar process as a mirror coat

[#231] Photosensitive lenses have one severe limitation you should always advise the patient of. That limitation is:

a) Polycarbonate photo lenses are very heavy
b) Photosensitive optics are poor and may lead to complaints
c) Glass reduces the UV that gets to the lens so they are a poor choice for driving behind a windshield
d) The time it takes for a photosensitive lens to darken and lighten is too long to be useful

[#232] Tinting lenses can be done with a variety of colors on most lenses. (NOTE: When combined with ARC add 15 to 20% to the tint as ARC will lighten it.) Which statement is **NOT** correct?

a) Tinting reduces light transmission
b) The wide variety of colors offers many cosmetic choices
c) Tinting a lens can damage a lens, is too time-consuming and is too expensive to be useful
d) Tints can be color-coordinated with the frame

[#233] Which statement is **NOT** correct?

a) Most of today's lenses typically have scratch resistance properties applied

b) The bonded scratch coat is "scratch-proof"

c) Proper cleaning should always be discussed with the patient regardless of scratch resistant properties that were applied

d) Originally scratch "coats" were sold as an add-on, but that practice is rarely seen today

Chapter Thirteen Practice Test

[#234] The right eye, in Latin, is:

a) Oculus Uterque

b) Oculus Dexter

c) Oculus Sinister

d) OU

[#235] The left eye, in Latin, is:

a) Oculus Uterque

b) Oculus Dexter

c) Oculus Sinister

d) OU

[#236] Referring to both eyes in Latin is:

a) Oculus Uterque

b) Oculus Dexter

c) Oculus Sinister

d) Oculus Uarice

[#237] Which letter points to the label signifying the left eye?

a) B

b) C

c) D

d) E

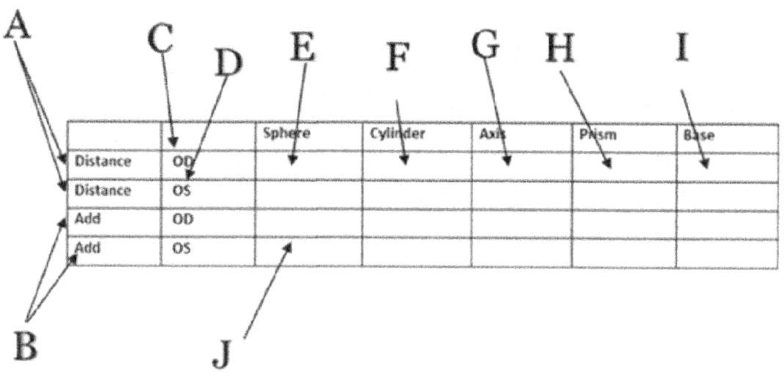

[#238] Which letter points to the prism direction box?

a) B

b) H

c) I

d) B

[#239] Which letter points to the box where the ADD would be written?

a) J
b) G
c) H
d) I

[#240] Which statement is **NOT** correct?

a) The sphere power on a toric lens is linked directly with the axis meridian written
b) If the cyl/axis box is blank, then the sphere power is present in all 180 meridians
c) If a cylinder power is written on the Rx, the cylinder power will always be 90 degrees from the axis number written on the Rx (irregular astigmatism will be the exception, but is extremely rare)
d) The power written in the axis box will always be the total power of the cylinder meridian

[#241] On Rx #1, what is the total power for the right eye at axis 45?

a) -5.00
b) -6.00
c) -1.00
d) +1.00

RX #1

		Sphere	Cylinder	Axis	Prism	Base
Distance	OD	-5.00	-1.00	45		
Distance	OS	-5.00	-1.00	135		
Add	OD					
Add	OS					

RX #2

		Sphere	Cylinder	Axis	Prism	Base
Distance	OD	-6.00	+1.00	135		
Distance	OS	-6.00	+1.00	45		
Add	OD					
Add	OS					

[#242] On Rx #1, what is the total power for the right eye at axis 135?

a) -5.00
b) -6.00
c) -1.00
d) +1.00

[#243] On Rx #1, what is the total power for the left eye at axis 135?

a) -5.00

b) -6.00

c) -1.00

d) +1.00

[#244] On Rx #1, what is the total power for the left eye at axis 45?

a) -5.00

b) -6.00

c) -1.00

d) +1.00

[#245] On Rx #2, what is the total power for the right eye at axis 135?

a) -5.00

b) -6.00

c) -1.00

d) +1.00

[#246] On Rx #2, what is the total power for the right eye at axis 45?

a) -5.00

b) -6.00

c) -1.00

d) +1.00

RX #1

		Sphere	Cylinder	Axis	Prism	Base
Distance	OD	-5.00	-1.00	45		
Distance	OS	-5.00	-1.00	135		
Add	OD					
Add	OS					

RX #2

		Sphere	Cylinder	Axis	Prism	Base
Distance	OD	-6.00	+1.00	135		
Distance	OS	-6.00	+1.00	45		
Add	OD					
Add	OS					

[#247] On Rx #2, what is the total power for the left eye at axis 45?

a) -5.00

b) -6.00

c) -1.00

d) +1.00

[#248] On Rx #2, what is the total power for the left eye at axis 135

a) -5.00
b) -6.00
c) -1.00
d) +1.00

[#249] Regarding Rx #1 and #2, what would be the difference between two lenses if Rx #1 was surfaced on one lens and Rx #2 was surfaced on the other, both on the back surface?

a) Rx #1 would be weaker than Rx #2
b) Rx #1 would be stronger than Rx #2
c) Rx #2 would be close in power to Rx #1
d) Rx #2 would be identical to Rx #1

		Sphere	Cylinder	Axis	Prism	Base
Distance	OD	+1.25	-0.50	80		
Distance	OS	+1.50	-1.00	75		
Add	OD	+2.00				
Add	OS	+2.00				

[#250] Which statement is <u>**NOT**</u> correct?

a) The right eye sphere power is +1.25
b) The right eye ADD is +2.00
c) The right eye total power in the reading area is +3.25-0.50 x 80
d) The right eye total power in the reading area is +2.00-0.50 x 80

[#251] What would the right eye Rx (+1.25-0.50 x 80 ADD +2.00) convert to for SV near-only glasses?

a) +3.25-0.50 x 80
b) +3.50-1.00 x 75
c) +2.00-0.50 x 80
d) +3.50-1.00 x 75

[#252] What would the left eye Rx (+1.50-1.00 x 75 ADD +2.00) convert to for SV near-only glasses?

a) +3.50-1.00 x 75
b) +3.25-0.50 x 80
c) +2.00-1.00 x 75
d) +3.25-0.50 x 80

[#253] What would the right eye Rx (+1.25-0.50 x 80 ADD +2.00) convert to for SV intermediate glasses?

a) +3.25-0.50 x 80
b) +3.50-0.50 x 75
c) +2.25-0.50 x 80
d) +2.50-1.00 x 75

		Sphere	Cylinder	Axis	Prism	Base
Distance	OD	+1.25	-0.50	80		
Distance	OS	+1.50	-1.00	75		
Add	OD	+2.00				
Add	OS	+2.00				

[#254] What would the left eye Rx (+1.50-1.00 x 75 ADD +2.00) convert to for SV intermediate glasses?

a) +3.25-0.50 x 80
b) +3.50-1.00 x 75
c) +2.25-0.50 x 80
d) +2.50-1.00 x 75

[#255] Which statement is **NOT** correct?

a) The most common patient condition that prism corrects is "double vision" (or strain caused by the muscles working too hard to fuse images)
b) Prism may be split primarily for cosmetic and balancing purposes
c) Typically with vertical prism, bases should be in opposite directions (up/down, down/up)
d) Typically vertical prism bases should be in the same direction (up/up, down/down)

[#256] An Rx states 10.00 diopters of base down prism in OD. What is the best prism split that maintains the integrity of the Rx to achieve weight balance and better cosmetics?

a) R 10 BD, L zero
b) R zero L 10 BU
c) R 5 BD, L 5 BU
d) R 7 BD, L 3 BU

[#257] Which statement is **NOT** correct?

a) If the Rx states OD 5 diopters base in, then OD 2.5 base in and OS 2.5 base in would be a good split prism configuration
b) If the Rx states OD 5 diopters base in, then OD 2.5 base in and OS 2.5 base out would be a good split prism configuration
c) If the Rx states OD 5 diopter base in, an equivalent configuration would be OS 5 diopters base in
d) The usefulness of a prism is its ability to bend light towards the base to make the image appear to have moved towards the apex

Chapter Fourteen Practice Test

[#258] The most critical measurement, for an optician, is:

a) Prentice's Rule

b) Snell's Law

c) Patient pupillary distance

d) Eye size

[#259] The difference between a DPD and a NPD is:

a) They are interchangeable for opticians; there is no difference

b) DPD is used with horizontal astigmatism; NPD is used with vertical astigmatism

c) The DPD measurement simulates the patient looking past 20 feet while the NPD measurement simulates the patient looking at something within arm's reach (18 to 26 inches)

d) DPD is for the dominant eye and NPD is for the non-dominant eye

[#260] Which statement is **NOT** correct?

a) With a pupilometer you may measure both the DPD or the NPD

b) The pupilometer reflects light off the patient's cornea to facilitate accurate measurement

c) The distance between the OCs in two lenses (in a pair of glasses) is unrelated to the PD

d) The NPD measures the patient's eyes when a converging gaze measurement is needed

Chapter Fifteen Practice Test

[#261] The optical center or OC is:

a) Always on the 180 line
b) At the midway point of the vertical meridian
c) The point where light is not refracted
d) At the geometric center of the edged lens

[#262] Which statement is **NOT** correct regarding Prentice's Rule?

P = prism in diopters
C = decentration in mm
D = lens power in diopters

P = (C)(D)/10

a) Use Prentice's Rule to determine how much prism is created when the OC is placed at a point that does not line up with the patients PD
b) In some cases, (typically stronger Rxs) OC decentration away from the PD can be used to create "wanted" prism without special lens fabrication (i.e. using a stock blank)
c) To create wanted base up prism in a lens you could decenter the OC the appropriate amount along the 180 line
d) Prism created by decentering is determined by multiplying the decentration in millimeters by the diopter power of the lens. This number is then divided by ten to determine the amount of prism created by that decentration

[#263] A 5 mm error on -3.50 diopter lens creates how much unwanted prism?

a) 1.25
b) 1.75
c) 2.75
d) 3.50

[#264] A 10 mm error on a -3.50 diopter lens creates how much unwanted prism?

a) 1.75
b) 2.75
c) 3.50
d) 4.25

[#265] The patient's DPD is 60 mm. The Rx is -5.00 OU. The lens OCs are 68 mm apart. How many diopters of prism is created?

a) Two
b) Four
c) Six
d) Eight

[#266] From question #265, what is the direction of the prism created?

a) IN
b) OUT
c) UP
d) DOWN

[#267] The patient's DPD is 72. The Rx is -4.75 OU. The glasses were manufactured with the OCs 62 mm apart. How much prism is induced and in what direction?

a) -4.75 base out
b) -4.75 base in
c) -4.75 base up
d) -4.75 base down

[#268] The patient's DPD is 57. The Rx is +4.75 OU. The glasses were manufactured with the OCs 67 mm apart. How much prism is induced and in what direction?

a) -4.75 base out (BO)
b) -4.75 base in (BI)
c) -4.75 base up (BU)
d) -4.75 base down (BD)

[#269] The distance between your patient's eyes is 32/32 mm. You inspect their new glasses (-7.00 OU) and discover the OCs are 30/30 mm apart. How much prism, and in what direction, is created?

a) 2.8 prism diopters base up
b) 2 prism diopters base out
c) 1.4 prism diopters base out OU
d) 2.8 prism diopters base in

[#270] The monocular PD of your patient is 33.5/32.0 mm. You inspect their new glasses (-7.00 OU) and discover the OCs are 30/30 mm apart. How much prism, and in what direction, is created in each eye?

a) OD 2.45 BI : OS 1.40 BO
b) OD 2.45 BO : OS 1.40 BI
c) OD 2.45 BO : OS 1.40 BO
d) OD 2.45 BI : OS 1.40 BI

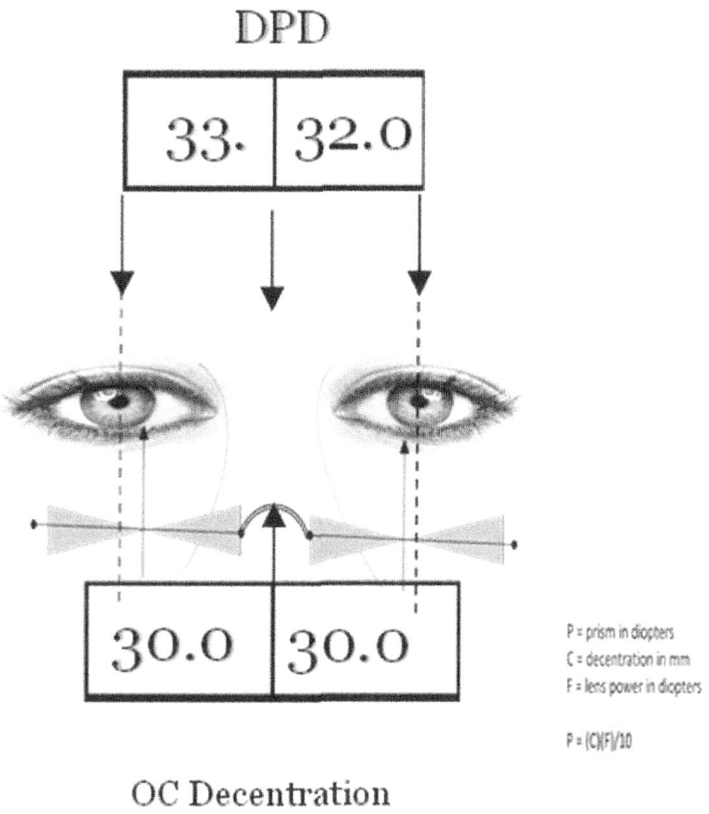

[#271] The monocular PD of your patient is 33.5/34.5 mm. You inspect their new glasses (-6.50 OU) and discover the OCs are 37.5/36 mm apart. How much prism, and in what direction, is created in each eye?

a) OD 2.60 BI : OS 0.98 BO
b) OD 2.60 BO : OS 0.98 BI
c) OD 0.98 BO : OS 2.6 BO
d) OD 2.60 BI : OS 0.98 BI

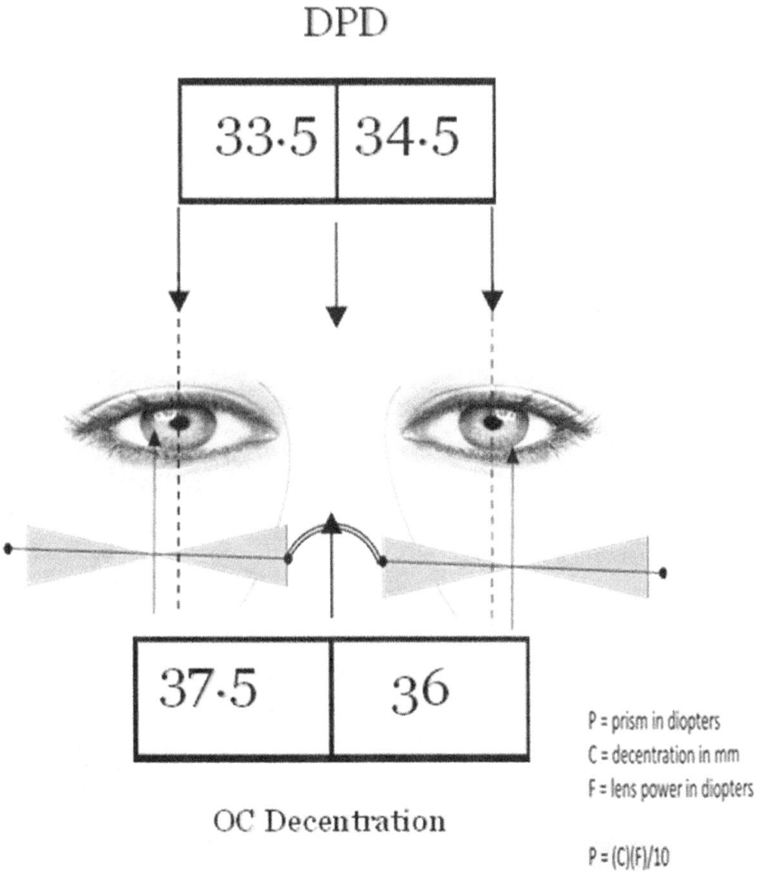

[#272] The monocular PD of your patient is 29.0/28.5 mm. You inspect their new glasses (-2.50 OU) and discover the OCs are 32/26 mm apart. How much prism, and in what direction, is created in each eye? Additionally, what is the total binocular prism?

a) OD 0.75 BI : OS 0.625 BO – total 0.125 BI

b) OD 0.75 BO : OS 0..625 BI – total 0.125 BO

c) OD 0.375 BO : OS 0.75 BO – total 1.12 BO

d) OD 0.375 BI : OS 0.75 BI – total 1.12 BI

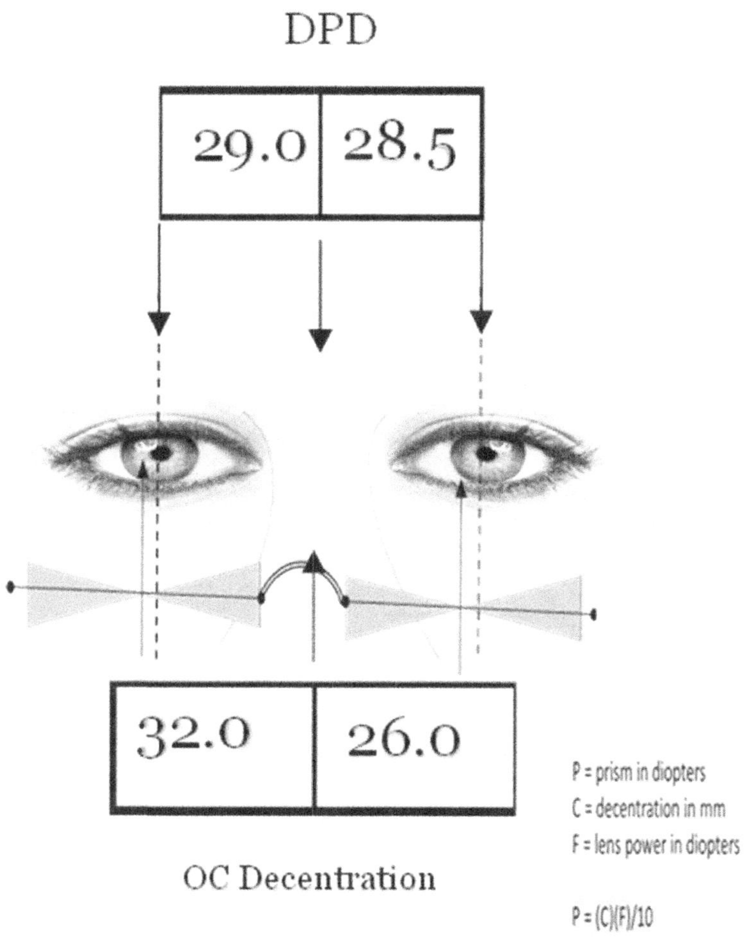

[#273] The monocular PD of your patient is 29.0/28.5 mm. You inspect their new glasses (+4.50 OU) and discover the OCs are 32/26 mm apart. How much prism, and in what direction, is created in each eye? What is the total prism?

a) OD 1.35 BI : OS 1.80 BI – total 3.15 BI

b) OD 1.35 BO : OS 1.125 BI – total 0.225 BI

c) OD 1.35 BO : OS 1.125 BI – total 0.225 BO

d) OD 1.80 BO : OS 1.35 BO – total 3.15 BO

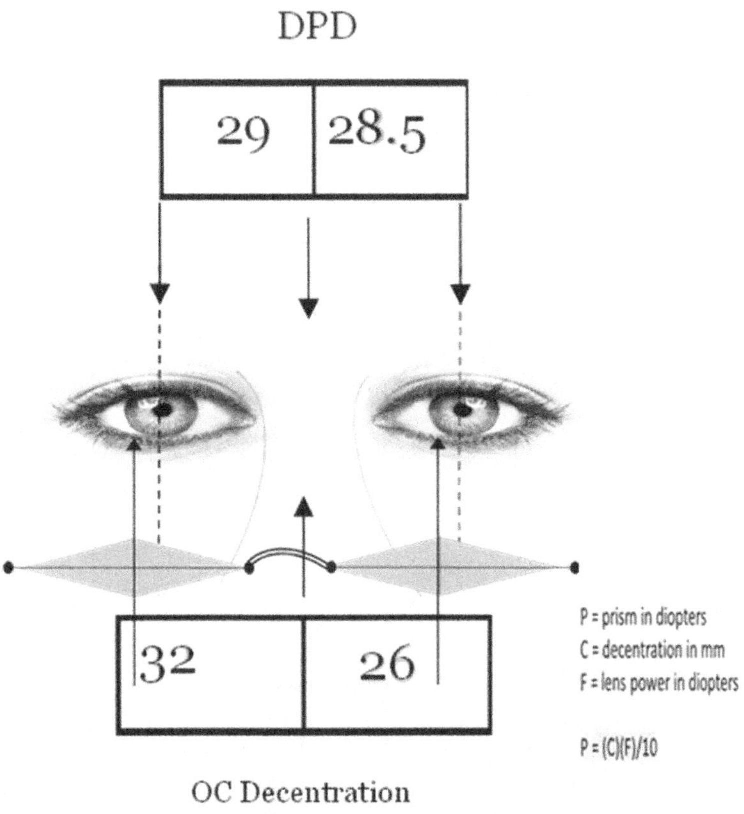

[#274] You inspect a patient's new glasses (-4.50 OU) and discover the OCs are 4 mm apart vertically. How much prismatic vertical imbalance has been created?

a) 0.9

b) 1.8

c) 2.7

d) 3.6

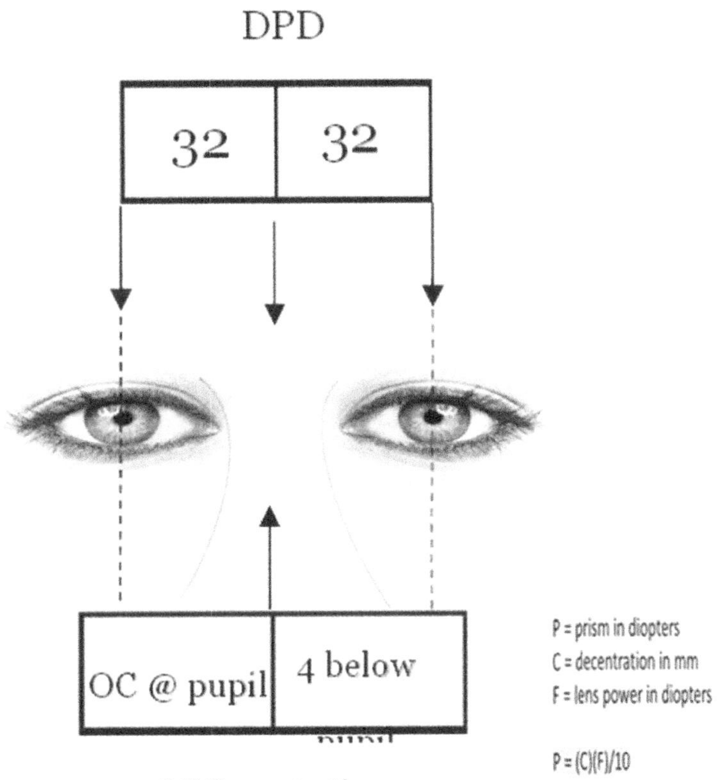

OC Decentration

[#275] Which statement is **NOT** correct regarding axis power determination (toric lens)?

a) The sphere power is along the meridian of the written axis
b) The power of the cylinder is along the meridian 90 degrees away from the written axis
c) The fact set described with answers "a" & "b" should only be used on a minus cylinder lens
d) The full power of the cylinder meridian is the sphere power added algebraically to the written cylinder power

[#276] What is the total power in the meridian of the sphere and the total power in the meridian of the cylinder in this lens (-1.00-1.00 x 180)?

a) -2.00 @ 180, -1.00 @ 90
b) -2.00 @ 90, -1.00 @ 180
c) -1.00 @ 180, -2.00 @ 90
d) -1.00 @ 90, -2.00 @ 180

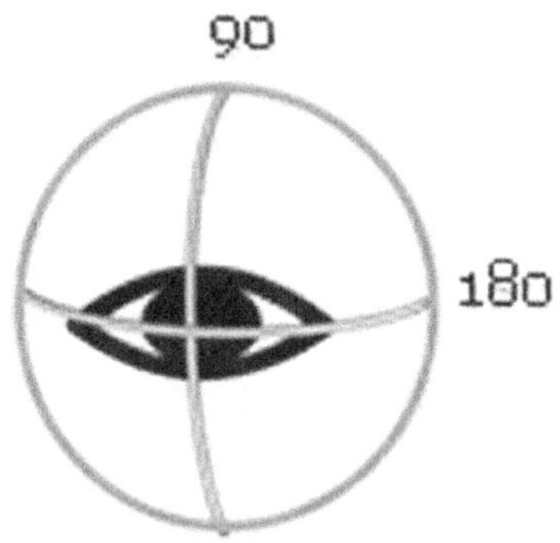

[#277] What is the total power in the meridian of the sphere and the total power in the meridian of the cylinder in this lens (-1.00-1.00 x 90)?

a) -2.00 @ 180, -1.00 @ 90
b) -2.00 @ 90, -1.00 @ 180
c) -1.00 @ 180, -2.00 @ 90
d) -1.00 @ 90, -2.00 @ 180

[#278] What is the total power of a lens (+3.50-1.00 x 135) at the meridian 135?

a) +2.50 @ 135
b) +3.50 @ 135
c) 135
d) +4.50 @ 45

[#279] What is the total power of a lens (+3.50-1.00 x 135) at the meridian 45?

a) +3.50 @ 45
b) +2.50 @ 45
c) 45
d) +4.50 @ 135

[#280] Which statement is **NOT** correct regarding the power on the 180 line of any prescription?

a) When the prescription is not at axis 180 or 90 you can still determine the power on the 180 line
b) Knowing the power on the 180 will assist understanding of the effect of decentration on the 180 line and when wanted or unwanted prism is present
c) The power on the 180 line is the sphere power plus the portion of the cylinder power that lies between the written axis and the 180 line
d) When the cylinder is 45 degrees away from the 180 line then 25% of the cylinder will be added to the sphere to determine the power on the 180

[#281] What is the power on the 180 line given this Rx (-4.25-3.25 x 60)?

a) -6.42 @ 60
b) -6.42 @ 180
c) -5.25 @180
d) -5.37 @ 180

$P = C\,(d/90) + S$

P = Power on 180
C = Cylinder Power
d = Cyl degrees to 0 or 180, less than 90
S = Sphere Power

[#282] What is the power on the 180 line given this Rx (+4.25-3.25 x 60)?

a) +5.32 @180
b) -5.32 @ 180
c) +3.17 @180
d) +2.08 @ 180

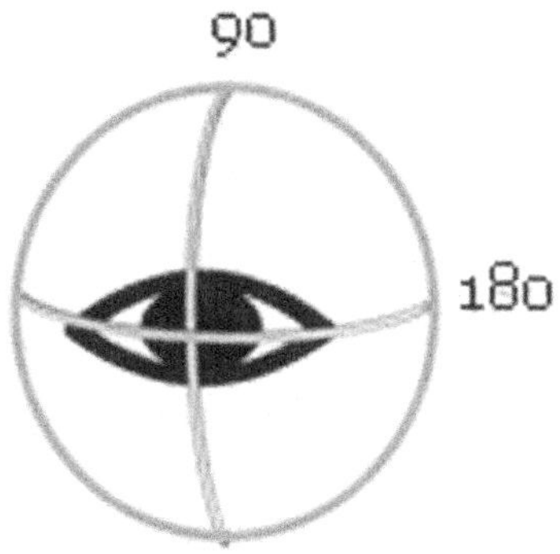

[#283] If this lens (**OD** -8.00-4.00 x 110) is ordered with a PD of 33, but was glazed into the frame with the OC at 36; how much prism, if any, was created and in what direction?

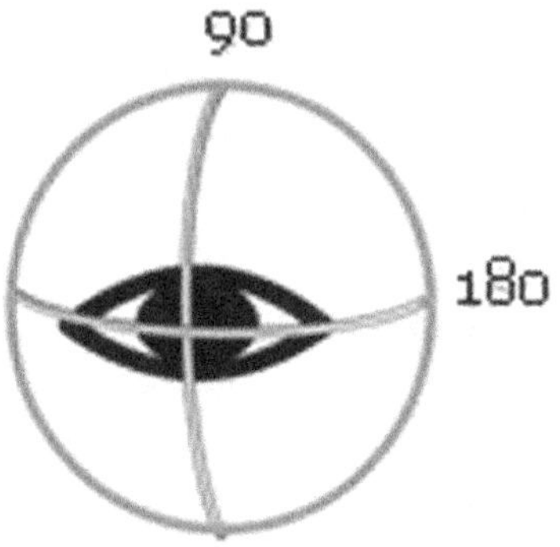

a) No prism created
b) 3.3 BI
c) 3.3 BO
d) .78 BI

Chapter Sixteen Practice Test

[#284] This illustrates measuring the:

a) A
b) B
c) Geometric Center (GC)
d) Effective Diameter (ED)

[#285] This illustrates measuring the:

a) A
b) B
c) GC
d) ED

[#286] This illustrates measuring the:

a) A
b) B
c) GC
d) ED

[#287] This illustrates the placement of:

a) A
b) Distance between the lenses (DBL)
c) GC
d) ED

[#288] This illustrates measuring the:

a) A
b) Patient PD (PPD)
c) Frame PD (FPD)
d) DBL

[#289] The DBL is synonymous with:

a) Eye size
b) B
c) Bridge
d) Box system

[#290] The MBS is:
a) Minimum B scale
b) Maximum blank size
c) Minimum blank size
d) Modified bridge size

[#291] This formula (FPD –PPD)/2 + ED) is used to calculate:
a) DPD
b) MBS
c) Eye size
d) Temple length

[#292] If the FPD equals 70, the PPD equals 60, the ED equals 49, the MBS is:
a) 44
b) 54
c) 75
d) 60

[#293] If the FPD equals 76, the PPD equals 62, the ED equals 58, the MBS is:
a) 49
b) 54
c) 75
d) 65

Chapter Seventeen Practice Test

[#294] Which statement is **NOT** correct?

a) The lensometer measures the focal length of the lens and converts to diopters on the power wheel
b) It is a best practice to "zero out" or calibrate the eyepiece before use
c) Failure to calibrate the eyepiece will typically result in generating insignificant errors
d) The lensometer can also be called a vertometer or a focimeter

[#295] The movable portion of the lensometer that can level and fix a lens or frame in position is called the:

a) Lens marker
b) Reticle
c) Table and lens holder
d) Target

[#296] When the need arises to visible mark the OC we employ the:

a) Lens marker pins
b) Reticle
c) Table and lens holder
d) Target

[#297] When the OC is properly centered on the table and the power correctly indicated, but the "in focus" target is displaced off-center, that indicates:

a) Reticle is out of focus
b) Table is in the wrong position
c) Prism in the lens
d) Lens marker in the way

[#298] The number of pins on the lens marker is:1

a) 2
b) 3
c) 4

[#299] The narrow lines on the target indicate:

a) Proper ink pin location
b) Cylinder power
c) Sphere power
d) Total power

[#300] The power difference between the focused narrow lines and the focused wide lines indicates:

a) Total power
b) Power of sphere
c) Amount of cylinder
d) Cylinder power

[#301] The orientation of prism can be manipulated with the:

a) Prism ring
b) Power wheel
c) Prism target
d) Reticle

[#302] The narrow lines come into focus at +3.00 @ 180 and the focused wide lines read +1.00 on the power wheel. The Rx is:

a) +1.00+3.00 x 180
b) +3.00+1.00 x 180
c) +3.00-2.00 x 180
d) +1.00-2.00 x 180

[#303] The narrow lines come into focus at -3.00 @ 75 and the focused wide lines read -6.75 on the power wheel. The Rx is:

a) -3.00-3.75 x 165
b) -3.00-3.75 x 75
c) -3.75-3.00 x 75
d) -3.00-6.75 x 75

[#304] The narrow lines come into focus at -5.50 @ 45 and the focused wide lines read -1.50 on the power wheel. The Rx in minus cylinder form is:

a) -5.50+4.00 x 45
b) -1.50-4.00 x 45
c) -4.00-1.50 x 135
d) -1.50-4.00 x 135

Chapter Eighteen Practice Test

Access The **Vision Council downloadable/viewable Quick Reference Guide** to view Ophthalmic ANSI Standards. These queries come from that document which was not included in this book due to copyright restrictions.

[#305] When measuring the distance power which type of lens has the greater room for error?

a) SV
b) Flattop
c) PAL
d) Myodisc

[#306] What is the greatest allowable error on a -10.00 spherical lens?

a) +/- 0.50
b) +/- 0.30
c) +/- 0.20
d) +/- 0.13

[#307] What is the greatest allowable error on a -15.00 spherical lens?

a) +/- 0.50
b) +/- 0.30
c) +/- 0.20
d) +/- 0.13

[#308] What is the greatest allowable error on a -1.00 spherical lens?

a) +/- 0.50
b) +/- 0.30
c) +/- 0.20
d) +/- 0.13

[#309] What is the greatest allowable axis location error on -5.00 cyl?

a) +/- 14 degrees
b) +/- 7 degrees
c) +/- 3 degrees
d) +/- 2 degrees

[#310] When a lens falls just slightly outside ANSI tolerances

a) Remake the lens
b) Get a second opinion
c) Pass the lens
d) Let the patient decide if it is good enough

[#311] Upon inspecting a pair of SV -3.00 OU glasses you note a 1.5 mm difference in the vertical OC position. How much vertical imbalance does that create and does it pass or fail?

a) 0.33, pass
b) 0.45, pass
c) 0.33, fail
d) 0.45, fail

[#312] A lens was prescribed at plano -0.25 x 45 and cannot be turned in the frame. Which lens below must be rejected?

a) Lensometer reads -0.12-0.25 x 59
b) Lensometer reads +0.12-0.12 x 31
c) Lensometer reads plano -0.25 x 60
d) Lensometer reads +0.12-0.25 x 35

[#313] When the doctor specifies a base curve it must be within:

a) +/- 1.25
b) +/- 1.00
c) +/- 0.80
d) +/- 0.75

[#314] The tolerance on the proper orientation of polarization is +/-:

a) 3 degrees
b) 5 degrees
c) 7 degrees
d) 9 degrees

[#315] On a segmented multifocal the allowable difference between the height of the segs is:

a) 1 mm
b) 1.25 mm
c) 1.50 mm
d) 2.0 mm

[#316] A +8.00-4.00 x 40 PAL may have a cylinder power error of less than:

a) 0.40 diopters
b) 0.30 diopters
c) 0.25 diopters
d) 0.20 diopters

Bonus Questions

These ABO-relevant concepts have been gleaned from various ABO discussions available on the interwebs. They are assorted topics.

[#317] Aniseikonia typically requires the use of:

a) Flatter base curves
b) Iseikonic lenses
c) Diverging lenses
d) Occluded lens

[#318] Prism is used when the eyes are misaligned. This condition is:

a) Strabismus
b) Retinopathy
c) Directional Rectus
d) Glaucoma

[#319] An emmetropic individual:

a) Will need a minus correction
b) Will need a plus correction
c) Will not need any correction
d) Will require prism

[#320] An anisometropic individual:

a) Will have a difference greater than two diopters between his or her eyes causing discomfort from the image size differences
b) Will need oblique prism prescribed
c) Has 20/20 vision
d) Is resistant to wearing progressive lenses

[#321] An ametropic individual:

a) Is the opposite of an emmetropic individual
b) Has perfect distance vision
c) Has double vision
d) Cannot wear glasses

[#322] An antimetropic individual:

a) Has perfect distance vision
b) Cannot wear glasses
c) Has anisometropia with one eye myopic and the other eye hyperopic
d) Has Against-the-Rule astigmatism

[#323] An optician can measure the amount of light transmitted through a lens with a:

a) Phoroptometer
b) Dioptometer
c) Spectrometer
d) Polameter

[#324] If you look through a plus lens and move it back and forth the image will:

a) Have a "scissors" motion
b) Move in the same direction as the lens
c) Not move
d) Move in the opposite direction

[#325] The one word an optician should never use to describe any lens is:

a) Prescribed
b) Glazed
c) Shatterproof
d) Corrected

[#326] The most common form of strabismus is:

a) Myopia
b) Crossed eyes
c) Fresnel rejection
d) Hyperopia

[#327] There is no lens in the lensometer but the power wheel reads -0.37. What should you do?

a) Replace the eyepiece
b) Clean the eyepiece
c) Focus the eyepiece
d) Compensate for the difference

[#328] Which answer is the spherical equivalent of -7.00-6.00 x 180?

a) -10.00
b) -11.00
c) -12.00
d) -13.00

Answer Key for Practice Tests

1	c
2	d
3	c
4	a
5	c
6	b
7	a
8	b
9	c
10	b
11	d
12	d
13	d
14	b
15	a
16	c
17	b
18	c
19	c
20	d
21	c
22	a
23	c
24	d
25	b
26	d
27	c
28	c
29	a
30	d
31	d
32	d
33	b

34	c
35	c
36	c
37	a
38	d
39	c
40	d
41	d
42	c
43	b
44	d
45	a
46	c
47	d
48	c
49	c
50	d
51	b
52	d
53	c
54	c
55	d
56	a
57	c
58	d
59	c
60	a
61	c
62	d
63	d
64	c
65	c
66	b
67	c
68	a

69	d
70	c
71	b
72	c
73	a
74	b
75	b
76	c
77	c
78	a
79	d
80	b
81	d
82	c
83	b
84	a
85	c
86	a
87	d
88	c
89	d
90	d
91	d
92	d
93	d
94	c
95	b
96	a
97	d
98	c
99	a
100	c
101	c
102	d
103	d

104	a
105	a
106	d
107	c
108	b
109	d
110	c
111	c
112	a
113	c
114	d
115	a
116	a
117	d
118	d
119	d
120	d
121	d
122	a
123	d
124	c
125	c
126	a
127	c
128	a
129	d
130	a
131	a
132	b
133	a
134	d
135	c
136	d
137	d
138	b

139	c
140	b
141	c
142	c
143	a
144	c
145	a
146	a
147	c
148	b
149	d
150	c
151	d
152	a
153	b
154	d
155	c
156	c
157	a
158	d
159	c
160	c
161	a
162	b
163	b
164	c
165	d
166	b
167	a
168	c
169	a
170	a
171	a
172	b
173	a

174	c
175	b
176	c
177	d
178	b
179	c
180	b
181	a
182	d
183	a
184	d
185	c
186	b
187	b
188	a
189	a
190	c
191	d
192	b
193	d
194	c
195	d
196	c
197	d
198	a
199	a
200	d
201	a
202	c
203	a
204	c
205	a
206	c
207	d
208	a

209	d
210	a
211	c
212	c
213	b
214	c
215	c
216	b
217	a
218	b
219	c
220	e
221	a
222	f
223	b
224	d
225	c
226	a
227	a
228	c
229	c
230	c
231	c
232	c
233	b
234	b
235	c
236	a
237	c
238	c
239	a
240	d
241	a
242	b
243	a

244	b
245	b
246	a
247	b
248	a
249	d
250	d
251	a
252	a
253	c
254	d
255	d
256	c
257	b
258	c
259	c
260	c
261	c
262	c
263	b
264	c
265	b
266	a
267	a
268	a
269	c
270	c
271	d
272	a
273	c
274	b
275	c
276	c
277	d
278	b

279	b
280	d
281	b
282	d
283	b
284	d
285	b
286	a
287	c
288	c
289	c
290	c
291	b
292	b
293	d
294	c
295	c
296	a
297	c
298	c
299	c
300	c
301	a
302	c
303	b
304	d
305	c
306	c
307	b
308	d
309	d
310	a
311	d
312	c
313	d

314	a
315	a
316	d

Bonus Questions

317	b
318	a
319	c
320	a
321	a
322	c
323	c
324	d
325	c
326	b
327	c
328	a

References

ABO-NCLE. (2025). National Opticianry Competency Examination (NOCE) content outline and test specifications. https://www.abo-ncle.org/ABO/BasicCert/ABOContent_TestSpecifications.aspx

Adobe Stock. (n.d.). Eye anatomy [Image].

Bjork, E. L., & Bjork, R. A. (2011). Making things hard on yourself, but in a good way: Creating desirable difficulties to enhance learning. Psychology and the Real World: Essays Illustrating Fundamental Contributions to Society, 2, 56–64.

Bjork, R. A. (1994). Memory and metamemory considerations in the training of human beings. In J. Metcalfe & A. Shimamura (Eds.), Metacognition: Knowing about knowing (pp. 185–205). MIT Press.

Brooks, C. W., & Borish, I. M. (2007). System for ophthalmic dispensing (3rd ed.). Butterworth-Heinemann.

Distance, V. (n.d.). Retrieved from Wikipedia: https://en.wikipedia.org/wiki/vertex

Dodgson, N. A. (2004). Variation and extrema of human interpupillary distance. Proceedings of SPIE - the International Society for Optical Engineering, 5291, 36–46. http://www.neildodgson.com/pubs/EI5291A-05.pdf https://doi.org/10.1117/12.529999

Dunlosky, J., Rawson, K. A., Marsh, E. J., Nathan, M. J., & Willingham, D. T. (2013). Improving students' learning with effective learning techniques: Promising directions from cognitive and educational psychology. Psychological Science in the Public Interest, 14(1), 4–58. https://doi.org/10.1177/1529100612453266

Gemini. (2026). Anatomy of the human eye diagram [Images].

Giancoli, D. C. (2014). Physics: Principles with applications (7th ed.). Pearson.

Halliday, D., Resnick, R., & Walker, J. (2013). Fundamentals of physics (10th ed.). Wiley.

Hecht, E. (2017). Optics (5th ed.). Pearson.

Helmenstine, A. M. (2025, June 10). The visible spectrum: Wavelengths and colors. ThoughtCo. https://www.thoughtco.com/understand-the-visible-spectrum-608329

Hewitt, P. G. (2014). Conceptual physics (12th ed.). Pearson.

Jalie, M. (2016). Principles of ophthalmic lenses (5th ed.). Association of British Dispensing Opticians.

Jalie, M. (2016). Principles of ophthalmic lenses (5th ed.). Association of British Dispensing Opticians.

Karpicke, J. D., & Blunt, J. R. (2011). Retrieval practice produces more learning than elaborative studying with concept mapping. Science, 331(6018), 772–775. https://doi.org/10.1126/science.1199327

Laramy-K. (2024). The ultimate ABO study guide v6.23. OpticianWorks. https://opticianworks.com/wp-content/uploads/2024/01/Laramy-K-OpticianWorks-Ultimate-ABO-Study-Guide-v6.23.pdf

Light. (n.d.). Retrieved from Wikipedia: https://en.wikipedia.org/wiki/Light

National Institute of Standards and Technology. (2024). The physics of light and color. U.S. Department of Commerce. https://www.nist.gov/pml/owm/metric-si/si-units

OptiBoard. (2024, April 1). I passed the ABO - my experience and advice! https://www.optiboard.com/forums/forum/optical-forums/general-optics-and-eyecare-discussion-forum/75294-i-passed-the-abo-my-experience-and-advice

Principles of Plane Reflection and Plano Surfaces. (n.d.). [Excerpt].

Roediger, H. L., & Karpicke, J. D. (2006). Test-enhanced learning: Taking memory tests improves long-term retention. Psychological Science, 17(3), 249–255. https://doi.org/10.1111/j.1467-9280.2006.01693.x

Scientific Diagram. (n.d.). Human eye section view [Images].

Serway, R. A., & Jewett, J. W. (2018). Physics for scientists and engineers with modern physics (10th ed.). Cengage Learning.

Textbook Illustration. (n.d.). Internal structure of the eye [Image].

Thompson, A., & Taylor, B. N. (2008). Guide for the use of the International System of Units (SI) (NIST Special Publication 811). National Institute of Standards and Technology. https://doi.org/10.6028/NIST.SP.811e2008

Walker, J. (2014). Fundamentals of physics (10th ed.). Wiley.

Walker, J. (2014). Fundamentals of physics (10th ed.). Wiley. Hewitt, P. G. (2014). Conceptual physics (12th ed.). Pearson

INDEX

M

N

O

P

R

S

T

U